T0175989

THE NEW FIRE

THE NEW FIRE

WAR, PEACE, AND DEMOCRACY IN THE AGE OF AI

BEN BUCHANAN AND ANDREW IMBRIE

THE MIT PRESS
CAMBRIDGE, MASSACHUSETTS
LONDON, ENGLAND

The MIT Press would like to thank the anonymous peer reviewers who provided comments on drafts of this book. The generous work of academic experts is essential for establishing the authority and quality of our publications. We acknowledge with gratitude the contributions of these otherwise uncredited readers.

This book was set in ITC Stone and Avenir by New Best-set Typesetters Ltd. Printed and bound in the United States of America.

Library of Congress Cataloging-in-Publication Data

Names: Buchanan, Ben, author. | Imbrie, Andrew, author.
Title: The new fire : war, peace, and Democracy in the age of AI / Ben Buchanan and Andrew Imbrie.
Other titles: War, peace, and Democracy in the age of artificial intelligence
Description: Cambridge, Massachusetts : The MIT Press, [2022] | Includes bibliographical references and index.
Identifiers: LCCN 2021013280 | ISBN 9780262046541 (Hardcover)
Subjects: LCSH: Artificial intelligence.
Classification: LCC Q335 .B795 2022 | DDC 006.3—dc23
LC record available at https://lccn.loc.gov/2021013280

10 9 8 7 6 5 4 3 2 1

For Mary
and
For Teresa

CONTENTS

INTRODUCTION

Fire begins with a tiny spark. The spark can come from two sticks rubbed together, a smoldering match left astray, an electrical fault, or any number of innocuous yet volatile triggers. The spark ignites and, with the right fuel, grows exponentially. What was once a nearly invisible flame can quickly turn into a small blaze, and then explode into a wildfire that rips across landscapes and destroys everything in its path. Fire's power comes not just from its heat and smoke but also from its acceleration in speed and force; unchecked, it will rage and rage until there is nothing left to burn.

And yet, for all the destruction that fire can bring, it is also the basis for civilization. From its earliest days, humanity's capacity to start, control, and stop fires was the prerequisite for warming caves and everything that followed. Many of the most consequential inventions—metal forging, steam propulsion, glassmaking, and electricity—arose from this ability. Some scientists theorize that even the development of the human brain itself depended on cooking meat over flames, which meant that food required less energy to digest.[1] In so many ways, human beings learned to harness fire's exponential power, using its might for good and taming its dangers.

Humanity has also wielded fire's destructive force. The Byzantine Empire used it to great military success, first during the siege of

Constantinople in 672 AD, and then in the centuries that followed. In battle, Byzantine troops shot a specially formulated compound at their enemies, one that would burn even when it came into contact with water.[2] Once the compound hit the target, the power of fire would kick in, torching enemy equipment and causing soldiers to flee. Since then, the flames of war have only become deadlier.

Could there ever be another force so productive and perilous, one so essentially defined by the exponential growth of its core components?

Welcome to the age of artificial intelligence.

In choosing the metaphor of fire to explain this age, we reject the more common and more hopeful assertion that AI is "the new electricity."[3] Electricity is everywhere in modern life, so ubiquitous and so utterly mastered by our society that we take its safety for granted. Without a second thought, we dwell and work side by side with the massive buzzing wires that crisscross our neighborhoods and empower our lives. Largely because of the professionalization of electricity generation, transmission, and usage, nearly every modern human interaction with electrical current is a beneficial one. AI will follow the same path, so the electricity metaphor implies, transforming society no less profoundly and no less positively.

While that may be true one day, for now this view is far too rosy. We have not come close to mastering AI as we mastered electricity.[4] Today, we encounter AI as our distant ancestors once encountered fire. If we manage this technology well, it will become a tremendous force for global good, lighting the way to many transformative inventions. If we deploy it too quickly and without adequate foresight, AI will burn in ways we cannot control. If we harness it to destroy, it will enable more powerful weapons for the strongest governments on Earth as they engage one another in a combustible geopolitical competition. That the frequent analogy to electricity denies this wide range of possible outcomes doesn't make us more secure, only less prepared.

Three sparks ignite the new fire: data, algorithms, and computing power. Today's AI systems use computing power to execute algorithms that instruct machines how to learn from data. Due to exponential growth in the size of data sets, the capability of algorithms, and the power of computers, the age of AI has brought advances that have stunned even some skeptics.

AI has demonstrated superb capability in areas ranging from speech and facial recognition to video generation, language translation, storytelling, and beyond. As a result of these advances, AI has come into our homes and our businesses. It empowers Siri and Alexa, provides recommendations as we navigate the web, helps steer our cars on the highway, and silently makes so many of the technological systems we use every day work better than before. Even more impressively, AI systems can now mimic decidedly human qualities, such as imagination and intuition. In rapid succession, these systems have reached milestones—including success at games that require fiendishly complex strategic planning and breakthroughs in some of the hardest science problems—that experts had once thought were more than a decade away. It is overwhelmingly likely that even more surprising and powerful AI advances will come in the years ahead.

AI's new capabilities are both marvels and distractions. Too often, the discussion of AI focuses on what the technology can do, overlooking the people who invent, refine, and deploy it. Commentators make breathless predictions about the arrival of superintelligence capable of exceeding humanity's cognitive capabilities, but far less often do they consider how AI interacts with geopolitics.[5] And yet history reminds us that, from chariots to cannons, from airplanes to atomic bombs, more important than the innovation itself is how and why people wield it.[6] Any inquiry into AI must focus on us—how we harness the new fire and for what purpose.

This book is about human choices. The decisions people in governments, companies, and universities make about AI will prove to be some of the most consequential of this young century, both technologically and geopolitically. AI compels us to ask anew some fundamental questions about humanity: How do we govern? How do we fight? And, perhaps most importantly, what do we fight for? We present three competing and sometimes overlapping views—the evangelists, warriors, and Cassandras—for what to do about this new technology.

The evangelists believe that humanity should tend the new fire for the benefit of all. Those who hold this view believe that AI can transform human civilization for the better, and worry that using AI as a weapon of war and geopolitics distracts from all the good it can do for civilization. They are often AI researchers and pioneers themselves, motivated by the desire to "solve intelligence" as a means of cracking some of the thorniest mysteries in science. At a time when scientific discovery and invention seem more difficult than ever before, evangelists want to build machines that can see, create, plan, and aid humanity in reaching its full potential. Just like the discovery of fire, discoveries in AI—or by AI—will unlock a new standard of human life and a new level of understanding. If we manage it well, the evangelists believe, the new fire's exponential force can propel us all forward and offer positive-sum benefits for everyone.

The warriors want to harness the new fire not only for science but also for national security. Just as nearly every military since the Byzantines has been ready and able to burn its enemies with one kind of flame or another, many strategists today think that AI can be useful in deterring and winning wars. They believe that few matters are removed from geopolitical competition, and that nations will act on AI as they do on any other global issue, such as climate change or pandemics: as self-interested players, usually competing and only sometimes cooperating with one another. Democracies in particular must be ready to wield the new fire, warriors believe, or they risk watching from the sidelines as autocracies use AI to centralize control at home and extend their influence globally; to be slow to adapt is to cede the advantage to others. The need to deploy AI for security and geopolitical advantage might be unsettling to contemplate, the warriors acknowledge, but it would be naive to ignore it. Sometimes the world is more zero-sum than we would like to admit.

The Cassandras fear the new fire. Named for the Trojan priestess whose prophecies were apt but ignored, they contend that AI is less useful and more dangerous than many proponents anticipate. While the warriors worry that democracies might be too slow to apply AI to national security, the Cassandras fear that nations might be too quick to do so. AI's intrinsic weaknesses concern them, as does its potential to be both powerful and inept; they warn that AI systems will catastrophically fail at moments of great importance. If deployed too quickly and without adequate foresight and ethical constraints, AI might be wildly destructive and negative-sum—not in the sense of a vengeful Terminator but rather in the emotionless might of a roaring blaze.

We use the views of the evangelists, warriors, and Cassandras to reflect divergent assumptions and varying points of emphasis, but at times the lines between these categories blur. While some people may hold the same view across many issues at the intersection of AI and geopolitics, others shift perspectives depending on the question. An evangelist can worry about the risks of AI, just as a warrior or a Cassandra might believe in the technology's potential benefit for the common good.

All three perspectives have their merits. AI is technically impressive, and it seems likely that the technology could assist with discoveries that will make life richer and better for much of humanity. At the same time, AI has already become a flashpoint between the United States and China, and between the forces of democracy and autocracy more broadly. Both superpowers invest billions of dollars each year in AI not only to advance humanity but also to gain an economic and military advantage over the other. This competition, like the new fire itself, is only accelerating. The warriors are right to want democracies to win, but the Cassandras are right to warn of a race to the bottom with deleterious effects on human rights, the reliability of the technology, and international security.

The interplay between the three views will determine what AI can do, who will benefit from it, and which consequences will befall us all. Representatives of each view will appear in the pages that follow. Some push AI's technological frontier forward, making breakthroughs that were once unfathomable. Others marshal these inventions to advance the interests of their nations, whether democratic or autocratic. Still others urge caution, warning that we are crossing ethical lines or creating unintended consequences. Sometimes the Cassandras may succeed with their efforts to minimize risk, but geopolitical competition and AI research will nevertheless continue to accelerate. AI will shape statecraft, and statecraft will shape AI.

AI AND STATECRAFT

First, we must explore the sparks of the new fire. Part I, "Ignition," examines the triad of data, algorithms, and computing power.[7] These chapters show how modern AI systems mark a notable change from previous decades of software engineering. With some kinds of early AI and other traditional forms of software, developers gave instructions to their creations and watched those machines carry them out by rote. With modern

AI, software developers instruct the system on how to learn, giving it data or environments from which to extract patterns and providing ample computing power. This different technical architecture creates the possibility of AI systems capable of making their own inventions and forming their own insights.

As these chapters show, the geopolitical implications of the evangelists' technical advances in data, algorithms, and computing power are inevitable, and the warriors are quick to spot them. Academic contests in computer image recognition, which draw on massive data sets, prompt governments to develop tools to automate imagery intelligence analysis. Algorithms that dominate at some of the most complex board and video games on Earth spur military investments and advances. The creation of ever-more powerful computer chips propels AI systems to new technological heights, yet also leads to fierce showdowns between rival nations over control of the supply chains for chip manufacturing equipment.

But for as much as AI can surprise us with its successes, it can also confound us with its failures. Today's AI systems are astoundingly sharp in some areas and mind-bogglingly inept in others. Worse, the systems fail in ways that exacerbate preexisting biases, give illusions of impartiality, provide little or no explanation for their decisions, depart dramatically from human expectations, reveal a total lack of understanding, and snowball into greater harms—all outcomes that portend significant negative geopolitical consequences. We close part I by exploring the technological roots of these failures, showing how they often emerge from the architecture of modern AI itself.

Part II, "Fuel," shows that, despite AI's many failings, its successes are too seductive for the warriors to resist. Nations are already putting AI to use in service of their security interests. Government planners believe that AI will inevitably have strategic implications, and they covet the advantages it bestows on their nations. In this global competition, it is not just governments that matter but also the private companies that invent cutting-edge technology. Yet the relationship between the private and public sectors is more contentious today than in the past, especially in the United States. Traditionally accustomed to leading the way in high-tech innovation, the Pentagon is now trying to catch up. It leans heavily, and with mixed success, on technology companies and academic researchers in its efforts to gain a military edge over other nations. In contrast, some strategists argue that China is more unified in its approach (if not always in its execution), drawing heavily on the products and employees of its domestic technology companies to tighten its autocratic grip.[8] Both nations are using AI to invent the future of war and peace.

Perhaps no invention in this geopolitical competition is more relevant or more controversial than lethal autonomous weapons. Warriors dream of an AI that transforms the battlefield, neutralizing incoming threats and targeting enemies as needed while still adhering to ethical principles. They imagine new tactics and capabilities, like automatic threat identification, drone swarm warfare, and advanced missile defense. They assert that too much human oversight will slow combat down, that only AI-enabled systems can react rapidly enough against AI-powered aggressors, and that military systems and machines must become even more capable.

These dreams are inching closer to reality. Many US military leaders view lethal autonomous weapons as key to managing the threat from China, a nation of rapidly increasing concern to American defense planners. In coming to this view, the Pentagon draws on an array of tests showing how lethal autonomous weapons would increase the strength of the US armed forces. China, meanwhile, says that it would support prohibiting fully autonomous weapons on the battlefield but that it is not opposed to their development or production—a position that seems untenable.[9] Experts in the human rights and technical communities urge caution to both nations, warning that AI won't be up to the job of killing justly.

Kinetic weapons are not the only way to do harm. Cyber capabilities are an essential tool of modern statecraft, and AI is poised to transform the digital landscape. AI systems already help plan, deploy, and thwart cyber operations, which have become only more potent and ubiquitous. Worse still, AI systems are themselves vulnerable to hacking and deception; they possess all of the regular weaknesses that plague other computer systems, plus new and vexing flaws. As geopolitical competition intensifies, warriors will compete to hack each other's autonomous systems, taking capabilities out of their creators' control and exploiting weaknesses to gain an advantage over others.

Even more insidiously, AI offers the prospect of automating propaganda and disinformation campaigns, adding more fuel to the new fire. AI can write text that seems genuine, generate videos that seem real, and do it all faster and cheaper than any human could match or detect. AI also shapes the terrain on which disinformation efforts unfold. It influences which stories appear on Facebook news feeds, which tweets show up on Twitter timelines, and which video pops up next on YouTube's autoplay. Major internet platforms have become arenas where worldviews clash every day, with malicious actors trying to coax corporate algorithms to make their messages go viral, and companies trying—and often failing—to use AI to keep disinformation and other forms of hate at bay.

Part III, "Wildfire," looks at how the flames can burn out of control. It focuses on how mitigating fear—and achieving security—is an indelible part of modern statecraft. But each nation's quest for its own security can threaten others and exacerbate fear, especially when new technology is involved. It is this fear that leads warriors to take AI out of research labs and incorporate it into weapon designs and war plans—including for nuclear conflict. Since arms control around AI is nascent at best, governments use AI to aid their intelligence efforts to gain peacetime advantages over others. Such fierce AI-enabled competition in peace and war only increases the potency of combat and the risks of inadvertent escalation.

When it comes to geopolitical competition, a worrying possibility runs throughout this book: that AI will do more for autocracy than it will for democracy. In this view, autocrats will simply be better able or more willing to use AI to control people, information, and weapons. Unencumbered by privacy laws, they will have access to more data. They will use their central planning systems to drive algorithmic research and build ever-faster computers, turning autocracy's biggest weakness—the need for centralization—into its greatest strength. AI will help entrench power within the state, aiding in surveillance and repression and compelling private companies to assist as needed. For national security, the argument goes, AI will also improve the tools in the autocrat's arsenal, such as lethal autonomous weapons, cyber operations, and disinformation. In autocracies, ethical concerns will be nothing but minor speed bumps.

Equally troubling is the growing consensus that democracies will struggle to adapt to the age of AI. If this consensus is accurate, privacy concerns, civil liberties, and ethics will hamper data collection and slow the adoption of technological advances—as at times should be the case, given the rights of citizens in democracies. Companies will continue to develop cutting-edge algorithms, but they will be used for increasing advertising effectiveness, improving product recommendations, or advancing medical treatments—not strengthening and securing the state. The will required for ambitious long-term public projects, including for developing advanced computing power, will never materialize in democracies consumed by political infighting. In this view, ethical concerns will make democratic governments hesitate to develop lethal autonomous weapons, even as autocracies' new capabilities, including hacking and disinformation operations, put democracies on the defensive. As a result, democracies might struggle in peacetime, lose battles in wartime, and fall behind autocracies in general. In the words of one of the most-quoted AI thinkers of our moment, the historian Yuval Noah Harari, "technology favors tyranny."[10]

Yet nothing is as inevitable as these forecasts suggest, and this dystopian scenario will arise only if we mismanage the new fire. When it comes to AI, democracies have advantages; dictatorships have flaws. For each subject of the chapters that follow—from technical matters like the three sparks of the new fire through geopolitical subjects like lethal autonomous weapons—democracies have the opportunity to forge a way of using AI that assures, rather than undermines, their ideals and their security. It will not be easy. Democratic governments will have to deploy AI in concert with both their allies and their own private sectors. They will have to wield it to defend themselves, guard against its use as a tool of disinformation, and yet respect the human rights and fundamental values at the core of the social contract. Even the competition with autocracies need not be all-consuming, and there may be ways to manage the spirals of insecurity that seem inevitable to so many.[11]

The evangelists, warriors, and Cassandras agree on one thing: for better or worse, the geopolitical decisions that democracies make today will govern the future of AI as surely as the technical decisions will. Much remains in the balance, and the stakes could not be higher. Enormous data sets, massively capable algorithms, and rapidly growing computing power have lit the flames. Democracies and autocracies are together already fueling the blaze, developing AI and deploying it in peace and in war. They will try to harness the heat and smoke to their benefit, sometimes with success and sometimes courting immense risks. Their choices, and the competition that ensues, will determine where the new fire will bring warmth and light, and where it will burn everything to ash.

I

IGNITION

1

DATA

HAL 9000 appears onscreen for only a few minutes of *2001: A Space Odyssey*, but the spaceship computer is considered one of the greatest film villains of all time.[1] So adept with language that it can understand human speech and even read lips, HAL wields great power by controlling the ship's mechanics, and eventually becomes an antagonist for the crew. In the actual year 2001, such a performance from AI was still science fiction—but two researchers at Microsoft were doing their small part to turn fiction into fact.

The computer scientists Michele Banko and Eric Brill studied a key part of AI known as natural language processing, or the science of developing computer programs that can recognize, interpret, and manipulate written or spoken words.[2] Plenty of such programs existed at the time, but none of them were very good. Banko and Brill wondered what might be holding these programs back. They found that most software engineers working on natural language processing fed their programs about one million words of text, which the programs used to learn the patterns of human language.

Compared with the trillions of words spoken by humanity every single day, one million struck Banko and Brill as a trivial number. They became curious about what would happen if natural language processing programs had access to a lot more data. To test their idea, they devised an experiment, choosing a fairly simple and standard natural language proc-essing task: when asked to fill in a blank in a sentence with one of two similar but distinct words, like "principal" and "principle," could the AI determine which was correct?

Banko and Brill tried this with four AI programs, and found that, interestingly, the programs all performed similarly at the task. Despite the fierce competition between software engineers striving to design the best computer code, the differences between the programs didn't actu-ally matter all that much for performance. What determined results, they realized, was the amount of data supplied to teach each program how humans write and speak. With the standard one million words of sample data, each program chose the right word around 82 percent of the time. When Banko and Brill increased the amount of sample data to ten mil-lion words, the exact same four programs did better, with three of them exceeding 90 percent. With one hundred million words, accuracy pushed 95 percent. And with a billion words, three of the algorithms selected the right word around 97 percent of the time.[3]

The experiment underscored a key, and counterintuitive, idea: in the age of AI, software performance sometimes wasn't about the code. It was about the data.

Banko and Brill were prescient. Their 2001 experiment anticipated a paradigm shift that would transpire fully only years later, one that would place data at the center not just of AI but of society. In 2008, *Wired* maga-zine published a paean to what data could do, proclaiming that it would "replace every other tool that might be brought to bear. Out with every theory of human behavior, from linguistics to sociology. Forget taxon-omy, ontology, and psychology. Who knows why people do what they do? The point is they do it, and we can track and measure it with unprec-edented fidelity. With enough data, the numbers speak for themselves."[4] In such a worldview, data carried both technological and geopolitical sig-nificance. Google's Peter Norvig, a prominent computer scientist, said in 2011, "We don't have better algorithms than anyone else; we just have more data."[5] Ginni Rometty, IBM's then CEO, spoke of data as the "next natural resource."[6] Similar ideas soon proliferated, as data storage costs dropped and the size of databases all over the world increased exponen-tially. "Data is the new oil" became the mantra; "China is the Saudi Arabia of data" was the corollary.[7] Access to vast stores of information could fuel both AI and autocracy.[8]

Amid all this hype, one question persists even years later: *Is* data actually the new oil—that is, a resource that will become an underlying driver of economic activity and a cause of geopolitical conflict? The centrality of data to modern life and global affairs is more often proclaimed than it is examined. Few people know, for example, that this omnipresent oil analogy originated not from careful technical study or detailed geopolitical analysis but from grocery-store marketing. Coined in 2006 by a customer loyalty expert and data scientist named Clive Humby, the phrase was introduced to capture what Humby saw as the future of customer relations.[9]

As future chapters will show, data is only part of the picture for AI and statecraft. The oil tagline is simple and neat; it is also misleading. Like oil, data is important, but that's where the similarities end. Data is not a broadly useful commodity but rather information that is often specific to a particular context; it is not a finite resource but a foundation. Data does indeed matter, yet not in the ways so many expect.

As Banko and Brill foreshadowed, and as this chapter explores, data is at the center of a paradigm shift in AI that ushered in a new age of progress. This paradigm shift made the evangelists hopeful about what might come next, caused warriors to dream of how data-driven AI could serve their national interests, and prompted the Cassandras to worry about the implications that would follow. The rise of data was the first spark that lit the new fire; it is the beginning of our story, not its end.

WHY DATA MATTERS

Gottfried Leibniz preferred logic to bloodshed. An eminent German mathematician, he discovered calculus in the seventeenth century and made a number of other fundamental breakthroughs. Unlike most mathematicians, however, he concerned himself with philosophy as well. Leibniz understood the consequences of philosophical, political, and theological strife. He wanted to find a better way.

Turning to the methods of rationality that governed his work in mathematics, Leibniz aspired to create a formal system of logic that would be capable of processing all of the information in the world and deriving insight—a system that would provide a trustworthy mechanism for interpreting data and resolving debates without fighting. It would go far beyond even math as complex as calculus and would be capable of answering a wide range of questions. Leibniz wrote that, if such a system existed, "There would be no more need of disputation between two philosophers than between two accountants. For it would suffice for them to take their pencils in their hands and say to each other: Let us calculate!"[10]

Leibniz's ambition, in some broad sense, was an early aspiration toward creating AI. His idea contained an important characteristic that would later become the foundation of modern computing: a structured and replicable method of processing information to draw conclusions. The mathematician who discovered calculus might have hoped that the entire world could be subject to the same kind of rigorous analysis, if only the right rules and structures could be found.

Leibniz's vision has yet to materialize. He was a pioneer in building mechanical computers capable of performing basic operations, such as addition and subtraction, but these computers were obviously nowhere near complex enough to resolve political and philosophical disputes. There is simply too much data in the world, and that data is too complex and too varied, for one single logical system to understand it all. No set of known rules or analytic structures can process everything. Politics is still more akin to philosophy than it is to accounting.

Still, digital computers are the successors to Leibniz's early constructions. Computer languages provide a way for humans to give instructions to machines in a form they can understand; these languages encode some of the messy information from the world into a logical system that can form the basis of replicable calculation and analysis. Every computer program is a series of complex but carefully defined commands for how the program should interact with the data that it receives and use that data to inform its conclusions.

Centuries after Leibniz's proposals, AI came into vogue early in the modern computer age. Organizers of a summer research project at Dartmouth University coined the phrase *artificial intelligence* in 1956 and spurred the beginnings of the field.[11] AI researchers hoped that the structures of computer programs would eventually become powerful enough to form their own kind of intelligence. Early AI pioneers devised a type of program called an expert system and fed these systems carefully crafted rules for making decisions—often expressed in the form of logical statements like "if x is true, then do y."

The rules depended on the task at hand and on human knowledge of how to accomplish it. If the system was supposed to help oil engineers ensure performance from a well, programmers crafted these rules with input from drilling experts; if the goal was to determine the optimal price for a particular service, experienced field sales representatives helped shape the machine's behavior. To some degree, this rule-based method of processing data worked for many purposes. By the 1980s, two-thirds of Fortune 500 companies employed expert systems as part of their daily work.[12]

In reality, these early attempts at AI were only as good as their rules. Because of their rigidity and dependence on human-crafted instructions, expert systems were no substitute for intelligence. For very narrow tasks, they worked well enough, but when required to confront ambiguity or demonstrate flexibility in more complex tasks, they failed. Even in the days when expert systems were viewed with the most optimism, any hope of matching humans' general capabilities seemed to many observers to be a very long way away.

A competitor to expert systems, first developed in the late 1950s, lurked in the background: machine learning. Machine learning pioneers aspired to give rise to AI in a new way, one that differed both from the potential of expert systems and from Leibniz's vision. Machine learning systems focus on data, not on rules; it was on machine learning systems that Banko and Brill performed their experiment. The systems do not depend on humans painstakingly encoding knowledge and insights in the form of if-then statements and other explicit instructions. Rather, they distill what they need to know from vast stores of information. It is this difference—learning as opposed to simple execution—that makes machine learning both powerful and threatening, capable of delightful invention and also devastating failure.

At the core of the most powerful modern machine learning systems is something called a neural network. The workings of these networks are not intuitive to most people. They sound a bit abstract and arcane, more like a plaything dreamed up by a bored mathematician than the kind of tool that actually would be useful to anyone. But the remarkable fact is that a lot of the time neural networks work. They can turn data into insight.

Represented with code, a neural network consists of lots of nodes, which are called neurons. The word *neuron* usually conjures associations with the brain and complex neuroscience, but in AI it has a far simpler meaning: a neuron is just a container that stores a value, such as the color of a pixel or the result of a mathematical equation. These neurons are typically grouped together in layers, which are themselves arranged in sequence. Neurons make up layers, and layers make up the network; the number and size of the layers depends on how complex the task of the network is, with more neurons yielding more capability. Neural network–based systems are referred to as deep learning systems because of how these layers are stacked alongside one another.

To illustrate how neural networks function, and why they matter for national security, consider imagery intelligence analysis. It is a task that every major military performs. It begins when a satellite, drone, or spy plane takes a picture of something of interest. Analysts then try to determine what is happening in the picture. Depending on the resolution of the camera, the image might be blurry and hard to make out, and analysts could debate various possibilities for what it might show. During the Cuban Missile Crisis, for example, American intelligence officers analyzed photos of the island landscape to confirm the presence of Soviet missiles.

The right neural network can help identify objects within images and resolve some of these and other analytical debates. A neural network might look at a blurry picture of a vehicle and determine if it is a military tank or a civilian jeep. To do this, the network's first layer of neurons initially receives the image, with each neuron in the layer corresponding to the color (in a numerical form) of each pixel in the picture.[13] From this initial input, the network will eventually reach its conclusion.

For that to happen, however, the data in the first layer of the neural network has to cascade to the next one. In the simplest network designs, every neuron in the first layer connects to every neuron in the second layer, and the strength of these connections varies between each pair of

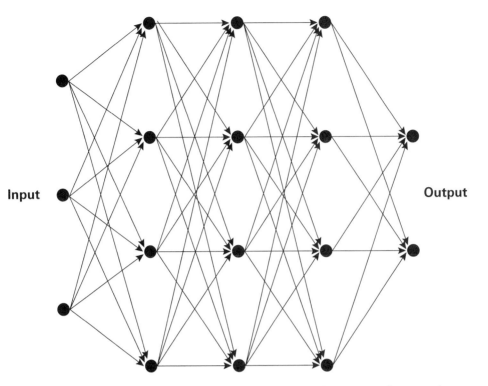

Input   **Output**

1.1 A neural network consists of many nodes, or neurons, that are typically grouped together in layers. Neurons from one layer usually connect to neurons from the previous and subsequent layers. Neural network–based systems are referred to as deep learning systems because of how these layers are stacked alongside one another.

neurons. The numbers contained within the second layer of neurons are determined by the numbers in the first layer—which represent the color of the pixels in the image—multiplied by the strength of each connection.[14] A higher value in the first neuron and a strong connection (represented by a larger number) will yield a higher value in the connected neuron.[15] Every neuron in the second layer is connected to every neuron in the third, which is in turn connected to every neuron in the fourth. The data flows in this way through the network, with the numbers inside neurons in one layer and the strength of their connections to the next layer determining the value of the next layer's neurons. Eventually the data, transformed by all of this math, reaches the last layer, which is called the output layer.

The output layer of a neural network has one neuron that typically corresponds to each possible answer. In the example of a neural network that determines if a vehicle is a tank or a jeep, the layer will have one neuron that represents a tank and one that represents a jeep. If, after all the cascading of data is done, the number within the tank neuron is larger than the number within the jeep neuron, the network has concluded that the image is likely a tank.

Essentially, the answer the output layer gives is the result of a very long mathematical equation. It begins with the image to be analyzed and then proceeds through each layer of the network, factoring into the equation the strength of each connection between neurons. The result at the end—the judgment about whether the image is a tank or a jeep—depends both on the image itself and on these connections. If the connections are tuned properly, the network will conclude that pictures of tanks are tanks and pictures of jeeps are jeeps.

There are far too many connections and other parameters in a neural network to set by hand, so humans do not determine the strength of these critical neuron-to-neuron links and, thus, the flow of information through the network.[16] Instead, many neural networks use a machine learning approach called supervised learning to set the strength of these connections. Rather than the explicit step-by-step reasoning of expert systems, classical computer programs, or Leibniz's formal logic, a supervised learning system depends on what AI researchers call training data.[17] The training data consists of examples that include the right answer—for instance, the training data for a simple system that distinguishes images of tanks from images of jeeps might include a thousand pictures of tanks labeled as tanks and a thousand pictures of jeeps labeled as jeeps. Banko and Brill showed that the more training data, the better the neural network performs.

All of this training data is typically provided to the neural network *before* it is deployed for use. Supervised learning systems learn all that they will ever know from the patterns in the training data, since they do not receive other instructions from their creators. No matter how obvious a fact is to humans, if it is not present as a pattern in the training data, a supervised learning system will never know it. Similarly, as chapter 4 will show, if the training data is not representative of real life, and indirectly includes patterns that are not emblematic of how the system is supposed to function, the system will have no way of identifying and counteracting those inadvertent biases. The centrality of data is perhaps the most important idea in all of supervised learning.

This dependence on data arises because a neural network using supervised learning is doing just that—*learning*. For each sample in the training data set, the network checks how it would have performed on that sample and compares it to the answer provided by humans. If, in one training example, the network incorrectly classifies a picture of a tank as a picture of a jeep, it then slightly adjusts the strength of connections within its network so that in the future it will be more likely to classify the picture and others like it as pictures of tanks. By changing the strength of the connections between neurons, the network changes the flow of data, and thus changes the numbers that arrive at the output layer that forms the system's answer.

The network makes these subtle tweaks to the strength of the connections between its neurons again and again and again as it processes all of the training data it is given. By looking at all of these images and comparing them to their human classifications as specified in the training data, the network learns which patterns make a tank a tank and which make a jeep a jeep. It learns to perform very well on the training data and on real-world examples that look like the training data.

When the network is fully trained, it will have learned all that it can from the data. It is ready to get to work on fresh data for which there is no right answer provided, such as a picture that has not yet been classified as containing a tank or a jeep. If this new data is similar to the data on which the network was trained—for example, if the new tank shares some of the same features as the tanks that the network has already seen, such as a long gun mounted on a turret—then a well-trained network will usually be able to recognize it correctly.

While the power of neural networks learning from data is now widely accepted in many fields, for decades there was great doubt. Years after Banko and Brill's work, it took new approaches to two old and vexing computer science problems—image recognition and image generation—to show the world what neural networks could do.

FIFTEEN MILLION IMAGES

Fei-Fei Li was born in Beijing, China, in 1976. When she was twelve, her father immigrated to New Jersey to start a life for the family in America. Four years later, Li and her mother followed. A day after her arrival in the country, the family car needed repair. Li's father, who spoke only basic English, brought her with him to the gas station and asked her to tell the mechanic to fix it; though she spoke little English herself, she communicated with gestures until the car was fixed. From then on, she realized, she had to become "the mouth and ears" of her parents.

Li was an academic star, especially in math. She so impressed one teacher at her local high school that he devised advanced classes for her during lunch hours. The same teacher loaned her family $20,000 to start a dry-cleaning business and helped her prepare for college. Two years after she arrived in the United States, Li earned a scholarship to Princeton University, where she studied math and physics. On weekends, while other students partied, she often commuted home to help out in the family business.[18]

A year after graduation, she moved to the famed California Institute of Technology to pursue a PhD that blended computer science with neuroscience, and eventually returned to Princeton in 2007 as a professor. In this early part of her career, Li worked on teaching computers to recognize images, a problem that had long bedeviled AI researchers. Image recognition, at least in some form, is a task that humans can carry out seemingly without thinking; just a few weeks after birth, an infant will recognize her mother's face.[19] But for decades, machines lacked the capacity to recognize what was in the collections of pixels that made up images. At best, they could adjudicate between a few options—perhaps differentiating a tank from a jeep—whereas an effective general image recognition system had to be able to differentiate a bike from a cloud from a woman from a tree from so much else. It seemed like there were just too many possibilities to teach a machine to recognize them all.

Li's idea was to create a gigantic collection of pictures. The database she named ImageNet contained photographs of a vast range of objects— buildings, Dalmatians, pickup trucks, toothbrushes, and more. Over time, it would grow to include more than fifteen million images, each of which needed a human operator to painstakingly caption and place it into one of thousands of categories so that it could eventually serve as training data for supervised learning systems.

Li paid Princeton students to do this labeling work, but the process still took too long, thanks to the sheer volume of images. A student suggested that Li try Mechanical Turk, an Amazon service that crowdsources low-cost labor to perform mundane jobs. Li gave it a shot and, after she built a verification mechanism that double-checked the workers' captions, she found that the labeling progressed much faster.[20]

In 2009, Li and her colleagues published a paper about ImageNet. It generated very little scholarly attention, and Li thought that an annual competition would boost awareness of the massive database. She realized that the vast collection of captions in ImageNet could be used to compare AI image recognition systems. If the caption offered by the AI matched the caption given by the humans who had first labeled the pictures, then the AI had succeeded in recognizing the image. Li named the contest the ImageNet Large Scale Visual Recognition Challenge, a mouthful of a name for what some would later call "the Olympics of computer vision."[21] Everyone who entered the competition would gain access to the same giant compilation of images. The winner would be the system that best extracted the patterns and insights from this data and used it to classify a subset of pictures selected by the competition organizers.

Designing such an image recognition system was exactly the challenge that Alex Krizhevsky and Ilya Sutskever had been seeking. The pair were PhD students at the University of Toronto, where they studied under Geoffrey Hinton, the neural network pioneer who will be discussed in chapter 5. In 2012, Krizhevsky and Sutskever decided to enter the ImageNet competition, and together they devised a system that came to be known as AlexNet.

AlexNet, like so many neural networks, became remarkably adept at its assigned task. Raw data went in one side, was processed by a bunch of neurons, and then a verdict—in this case an image caption—came out the other side. To accomplish this, AlexNet relied on 650,000 neurons arranged in eight layers, with more than sixty million total connections between neurons.[22] The strength of each of these links shaped how the data moved through the neural network and ultimately helped determine the answers that AlexNet gave when it was presented with an image. AlexNet used a supervised learning system to determine the optimal connection strengths between neurons and, as it did so, to learn the patterns in the ImageNet training data.[23]

Like most machine learning systems, AlexNet operated without direct human guidance. Krizhevsky and Sutskever did not tell it obvious facts about the images it would see, like the notion that four-wheeled vehicles were probably cars and two-wheeled vehicles were probably bikes. All they provided was ImageNet's millions of images and the labels on those images. By assembling the world's largest collection of human-captioned pictures, Li and her team had provided not just a means of evaluating image recognition systems but also the infrastructure for *improving* them. Networks that trained on Li's giant collection of labeled data would learn many more patterns than those trained on data sets that were smaller or contained less-precise captions. The information contained within ImageNet was a rising tide that would lift all image-recognition boats.

While neural networks and supervised learning had been applied to image recognition before, Krizhevsky and Sutskever made several improvements in the technology's efficiency and scale, enabling AlexNet to learn more patterns and extract more information from the captioned pictures in the ImageNet database. Among other improvements, AlexNet was bigger than a lot of other image recognition programs, with many more neurons and connections between these neurons. During its training process, AlexNet tweaked each of these connections time and again to best capture the patterns within the vast ImageNet collection and prepare the network to recognize images when the 2012 Olympics of computer vision began.

It worked. AlexNet beat out the competition, including many of the expert systems that were still prominent at the time. In the most common ImageNet scoring system—top-5 accuracy, in which the competitors got five guesses for each image—AlexNet succeeded around 85 percent of the time. The next best system managed only 74 percent.[24]

It was a triumph for Krizhevsky and Sutskever, for Li, and for AI progress more generally. AlexNet's success helped establish machine learning as the soon-to-be dominant paradigm in AI research. Olga Russakovsky, a Princeton professor and an organizer of the ImageNet project alongside

Li, said that "2012 was really the year when there was a massive break-through in accuracy, but it was also a proof of concept for deep learning models. . . . It really was the first time these models had been shown to work in context of large-scale image recognition problems."[25]

What happened next revealed a lot about AI and its place in the world. First, more researchers started using neural networks and the wheels of technological progress spun quickly, leading to rapid and widespread improvements in image recognition. Within two years, nearly every ImageNet competitor used a neural network architecture similar to AlexNet's. In each subsequent competition, these systems all got a lot better, in part because millions more images were added to the ImageNet database and in part because the programmers continued to improve how their neural networks learned. In the 2014 competition, the winning entry, known as GoogLeNet, scored better than 93 percent. GoogLeNet was not only more powerful and more efficient than all of the other competitors, but it was also on par with humans who engaged in the same image recognition task that year.[26]

Second, while the rapid progress in supervised learning was a technological marvel for everyone, irrespective of citizenship, it did not occur in a vacuum. The ImageNet competition very quickly took on geopolitical overtones, especially as an imperfect proxy for measuring nations' relative capabilities in AI research. The winning entry in the 2015 competition used further innovation to recognize almost 97 percent of images. Perhaps more interesting, though, was that the system's lead creator, Kai-ming He, had been educated in China and worked for Microsoft Research Asia in Beijing; it was the first time a China-based organization had won.[27] The next year, ImageNet's geopolitical stakes became more apparent: a research team from the Chinese Ministry of Public Security won one of the main components of the competition, a sign that government agencies had realized what machine learning could do for them. By 2017, the last year of the competition, more than half of the entering teams were based in China. That year, two researchers from a Beijing-based autonomous vehicle startup called Momenta devised the winning system.[28] ImageNet was the most famous competition in AI, and it had gone global.

These developments marked a rapid shift, in both technological and geopolitical terms. Neural networks with supervised learning systems and massive data sets had tackled a problem, image recognition, once thought to be manageable only by living creatures. In the span of just five years, these systems had gone from, at best, pretty obscure academic research projects to the spark for a new era of data-driven AI research. The successive ImageNet competitions and other supervised learning break-throughs showed that, with the right training data, machines could clas-sify text, images, sounds, and more. The potential applications were vast. After ImageNet's success, giant training data sets came into expanded use, teaching neural networks tasks ranging from detecting fraud in credit card transactions to forecasting the price of fine wine, which they could do more accurately than even the best wine connoisseurs.[29]

Li's work, and the work of other pioneers, kickstarted a rapid techno-logical transition toward supervised learning. It was a sea change that seemed primed to reach wherever there was data, including—as chapter 5 will show—the Pentagon and other military and intelligence organiza-tions. But as the ImageNet competition was taking off, another machine learning innovation would soon show that data could also serve a differ-ent and more creative purpose.

FROM IMAGES TO IMAGINATION

Ian Goodfellow had wanted to be the one to teach computers how to recognize images. For almost five years prior to the development of AlexNet, he had thought on and off about the problem, trying a variety of approaches.[30] None of them worked as well as what AlexNet and its successors managed to achieve. And so, as a PhD student in machine learning at the University of Montreal, he turned his attention elsewhere.

In 2014, Goodfellow and some friends were out at a bar in Montreal celebrating the graduation of a fellow doctoral student.[31] His friends had been trying to devise a system that did the opposite, in some respects, of what the ImageNet competition aimed to achieve: they wanted a com-puter to generate images rather than recognize them. Their plan was to design a program that could perform intricate statistical analysis on sam-ple photographs and then figure out what was in them—here was a chair, there was a table, and over there was a lamp. The code could then try to use that information to come up with new images of similar objects.

Goodfellow was skeptical. Image generation was a tricky problem. While supervised learning systems were by that time fairly proficient at using training data to learn how to recognize images, generating something new that fit the style of the data was more akin to imagination, and imagination seemed like a fundamentally human ability.

Goodfellow was imaginative himself. In high school, he had devised his own language with a phonetic alphabet, and enjoyed his creative writing and literature classes more than those in science and mathematics.[32] He hadn't really wanted to attend college but ended up going because his parents said they would pay for it and not for other pursuits. He picked Stanford because it would allow him to delay choosing a major until junior year and to leave and return with more flexibility than other schools; ever the literature student, he liked that John Steinbeck had attended Stanford intermittently as well.[33]

After his first year, Goodfellow decided to mimic the literary icon and take a break from college.[34] He moved to Scotland with two friends, animated by the goal of becoming a writer.[35] To pay the bills, he tried to land a day job anywhere he could, sending in applications for entry positions in factory work, fast food, and even forestry. No one hired him and, when his savings dwindled, he went back to Stanford. He took classes in neuroscience, chemistry, biology, and computer programming, thinking he would major in biomedical computation. Eventually, he decided to focus on AI but still encountered his share of setbacks; among them, Google rejected him for an internship.[36]

It is impossible to know if this circuitous and often nontechnical path helped Goodfellow grasp what it would take to develop a system capable of mimicking imagination. But in debating image generation over beers with his friends that night in Montreal in 2014, he hit upon an idea: the only thing better than one neural network would be two neural networks. Instead of his friends' plan to give explicit instructions to the system about which objects were present in an image and how they could be mimicked, Goodfellow wanted to kickstart the process with data. He would task one neural network with figuring out how to mimic some desirable data, such as photographs of people, and another with evaluating the work of the first network. He explained some early intuitions about how this could work to the assembled group at the bar.

They didn't buy it. When the celebration broke up, Goodfellow went home. In his own words, he could be "stubborn and argumentative," and the idea of two neural networks forming a single machine learning system lingered in his head.[37] While his girlfriend slept, he started to work then and there. His prototype showed strong promise on the first try; he later estimated that it had taken him only an hour to create it. By around midnight, he knew he was on his way to something big, and he began emailing colleagues to find collaborators to work on developing the idea further.[38] What Goodfellow eventually devised came to be known as a generative adversarial network, or GAN. GANs quickly became one of the most significant innovations in machine learning, and Goodfellow earned the moniker—in what can only pass for techie humor—of "the GANfather."[39]

A GAN is an unsupervised learning system, meaning it doesn't use the vast amounts of painstakingly labeled training data that supervised learning systems like AlexNet employ. Instead, the GAN uses data as a form of creative direction. It aims to create new information that resembles the data it is given, generating possibilities that mimic but do not duplicate a group of samples. It is this capacity for mimicking imagination and creation that makes GANs so special, even among other unsupervised learning systems.

At the core of a GAN's power is Goodfellow's deceptively simple idea of using unlabeled data to drive competition between two neural networks. The two networks' relationship is somewhat akin to that between a forger and an appraiser.[40] The first neural network, known as the generator, is tasked with devising something that looks like the data to be mimicked—it may be a picture of a human face, a video of a scene, or a musical composition.[41] The generator's first attempt will almost certainly be a spectacular failure, only minimally resembling the desired outcome. No matter how bad the output is, though, it moves along to the second neural network, the discriminator.

The discriminator receives the generator's product and tries to evaluate it. It also receives real-world data of the thing to be mimicked, such as a photo, video, or musical composition. For each item inputted into the discriminator, it must—like an expert examining a painting—render a judgment on whether the specimen is a fake produced by the first neural network or is in fact an authentic sample.

It is here that the true power of Goodfellow's design kicks in. Once the generator has produced its sample and the discriminator has made its judgment, both neural networks find out the result from the computer program that manages the GAN. The discriminator learns if its determination was correct; the generator learns if its sample was good enough to fool the discriminator. This feedback becomes the training data from which each network will learn. Similar to how AlexNet improved by assessing lots of pictures and adjusting its parameters to better capture what was in them, so too do the two neural networks in a GAN improve from the feedback they are given. The difference is that AlexNet's training data was labeled by humans, and the training data for the neural networks in a GAN results from the two networks' competitive process.

Through this process, the generator learns which outputs are most likely to fool the discriminator and therefore most closely resemble the samples of real-world data. The discriminator learns to spot small nuances in the specimens it examines so as to better determine which are samples of data from the real world and which are fakes that came from the generator. Just as iron sharpens iron, each neural network gets better at its task through competition with the other. Humans can sit on the sidelines and watch.[42]

Once the GAN is fully trained, it will feature a high-functioning generator and a very capable discriminator. This generator will be capable of mimicking imagination, or something like it. With some direction—the data that serves as samples—it is able to produce outputs that resemble but do not exactly replicate those samples. The success of Goodfellow's idea marked a pivotal advance in machine learning, showing how a single system could use data and two neural networks in concert. Machines could now try to do something else that had long seemed a uniquely human pursuit—creating something new—and they could do it without humans carefully labeling each piece of a large data set with the right answer.

Just as AlexNet sparked a quick succession of improvements in image recognition, GANs did the same for image generation. The results improved rapidly in a notable task: to depict a person who did not exist but who looked real. Goodfellow collected a sequence of pictures from research papers to show the rapid progress of successive systems in this

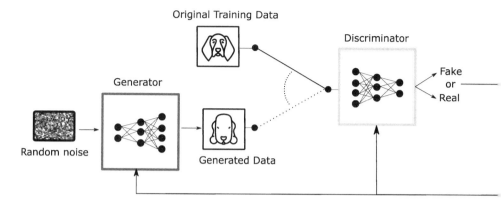

1.2 A GAN involves a generator and a discriminator. The generator produces an output designed to mimic some set of real-world samples, while the discriminator tries to determine if a given sample is from the generator or from the real world. Both neural networks receive feedback on their performance. This feedback becomes the training data from which they learn, with the generator becoming more adept at producing samples that seem real and the discriminator becoming more adept at determining which samples are real and which are from the generator.

area. In 2014, the best result from a GAN was blurry and unconvincing. In 2015, it was slightly sharper, but distorted and obviously fake, with eyes of uneven sizes. A year later, the best image was sharper still and quite convincing, but with odd eyebrows and a slightly asymmetrical lip. By 2017, state-of-the-art GANs were producing authentic-looking portraits, though they still sometimes made mistakes, such as depicting ears of atypically different heights or shapes. In 2018, as programmers devised better GANs with larger neural networks, that problem was largely resolved, and the faces the best GANs produced seemed strikingly credible.[43] This success went mainstream the following year with the creation of a website, www.thispersondoesnotexist.com, which shows an image of a convincing GAN-produced face each time a visitor opens the page.

With such capability for creation, GANs became more than a technical parlor trick. There are GANs now for nearly everything, including seemingly perfect videos of events that never happened.[44] The kinds of special effects that Hollywood spent millions devising are much more attainable with a GAN for a small fraction of the cost. The kinds of digital paintings that once had to be carefully done by hand can now be generated nearly

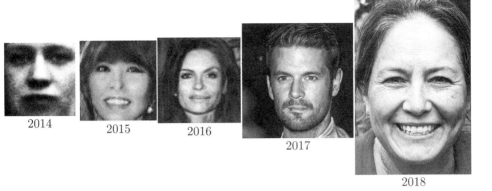

2014   2015   2016
                          2017

                                  2018

1.3 This timeline put together by Ian Goodfellow shows the rapid improvement in cutting-edge GAN outputs over time.

effortlessly by machine. By 2021, GANs and other generative technologies were so ubiquitous that another site, www.thisxdoesnotexist.com, was created to aggregate different generators and their creations, from pictures of cats to websites for startups and much in between.

It isn't just movie directors and artists who put GANs to work. Hackers and propagandists use GAN-created faces to make more convincing the online profiles from which they carry out social engineering operations.[45] Political leaders use GANs, too, including candidates who want to broaden their own appeal. In India, a politician deployed a GAN to deliver a message to supporters in a language that he did not speak.[46] The geopolitical implications of these and other uses of GANs are potentially significant but remain unclear, as chapter 8 will explore.

In these and other applications, the rapidly rising power of AI was obvious. In just a few years, large sets of data had helped make machine learning a force to be reckoned with. Supervised and unsupervised learning departed from centuries of conventional rules-based logic, offering new and often better ways to extract insight from that data. But the progress did not stop there. Machine learning soon began to shed some of its dependency on data. As Goodfellow was teaching machines to imagine, an ambitious young startup aspired to push the technological frontier still further, pioneering yet another type of machine learning system and attracting even more geopolitical attention.

2

ALGORITHMS

Before the Roman Empire and the Greek philosophers, before records of Chinese dynasties and Hindu kings, there were algorithms.[1] Babylonian mathematicians discovered that a series of well-organized and repeatable steps—an algorithm—could be devised to accomplish a particular task, beginning with some input and concluding with some output. If the input and the steps were the same each time, the output would also be the same. This idea was essential to order and structure in mathematics and all that was built upon that foundation. Two plus two will always equal four.

Profound as this insight was on its own, the crux of algorithmic thinking was that different sets of steps could be devised for different tasks, yielding different results even with the same inputs; the answer to the calculation two *minus* two will always be zero. The Babylonians crafted one process for multiplying numbers and an inverse one for dividing them.[2] Greek thinkers used more complex algorithms in their quest to find prime numbers. Islamic scholars in the ninth century developed innovative algorithms as they discovered algebra. In the 1800s, Charles Babbage and Ada Lovelace began to imagine what algorithms could do as part of general-purpose computing machines, which had not yet been invented. Lovelace in particular recognized that the numbers in algorithms didn't have to represent quantities but could represent more abstract concepts;

images or sounds, for example, could be converted to a numerical form and then given to a machine.[3]

The AI advances described in the previous chapter continued this history, first with expert systems and then with machine learning. While AlexNet and other data-centric supervised learning systems attracted interest in machine learning in 2012, and while GANs broadened the conception of what the technology could do, it was the subsequent algorithmic innovation described in this chapter that offered a second vital spark for the new fire. With ever more powerful and more efficient algorithms, AI could accomplish greater feats.[4]

Most significantly, algorithmic breakthroughs cemented the salience of AI to geopolitics. In changing what the technology could do, these algorithms also changed who cared about it. AI had once primarily been the sphere of technical engineers striving to maximize the performance of their systems. Now, it became the domain both of evangelists, who applied it to notable scientific problems, and of warriors within governments, who aimed to gain a strategic edge over rival nations. The gap between these two worldviews began to widen as it became more obvious just how powerful AI's algorithms would be.

ALGORITHMIC INTUITION

On May 11, 1997, the chess world champion Garry Kasparov sat down for the final game of his match with a computer program called Deep Blue. Deep Blue was an IBM-created algorithm that was perhaps the pinnacle of the expert systems paradigm. Grandmasters had advised the company on chess strategy, shaping the rules given to Deep Blue. The system used these thousands of lines of rules and insights to parse the many possible combinations on the chessboard and select the best move to play.

When Kasparov sat down to battle his automated opponent, he knew that it would be a historic day one way or the other. The six-game match was tied 2.5 points to 2.5 points, and this game would decide the winner. Many of the observers in the audience that day, including his mother, expected Kasparov to triumph. And everyone, especially Kasparov, knew the stakes: before an earlier match against Deep Blue, Kasparov had said he sought, on behalf of humanity, to "defend our dignity."[5] In its cover headline, *Newsweek* called the rematch "The Brain's Last Stand."[6]

Kasparov decided to open the game with the Caro-Kann Defense. It was an unusual opening, one often considered to be a weak choice in top tournament play, though one on which Kasparov was an expert. More confusing still was that Kasparov chose a seventh move that most top players shunned because it could give one's opponent the opportunity to sacrifice a knight in exchange for positional advantage. It appears Kasparov was trying to trick Deep Blue with his odd strategy. He was attempting to confound the expert system, thinking the algorithm would be unlikely to surrender one of its knights.

As it happened, Deep Blue was ready. Two months earlier, while preparing Deep Blue for the match, IBM's experts had happened to study games in which players had defeated the Caro-Kann Defense by sacrificing a knight. As a result, the computer knew more than Kasparov thought it did.[7] When the time came, Deep Blue correctly calculated that surrendering the knight in exchange for positional advantage made sense. Just eleven moves later, Kasparov resigned the game, stunning the assembled crowd in New York. He turned to his mother and raised his hands, seemingly in either frustration or surrender.[8] A computer algorithm had beaten a world chess champion at his own game.

The event garnered wide coverage. Some heralded it as a sign of ascendant artificial intelligence.[9] Others, including Kasparov, were much more dismissive, claiming that it was the giant set of human-provided insights, including the collection of openings, that had given Deep Blue a decisive advantage.[10] Still others criticized Kasparov, saying he was too passive in his play, too cute by half in his choice of the Caro-Kann, and too defeatist in his resignation.[11] No matter their views, nearly every expert acknowledged that, after chess, the next frontier was beating the world champion at Go, an even more complex game.[12]

For a computer to win at Go seemed nearly impossible. The game makes chess—which itself offers a vast number of possible move combinations—look downright simple. One expert described playing Go as "intensively contemplative, almost hypnotic . . . like putting your hand on the third rail of the universe."[13] Players alternate placing stones on a nineteen-by-nineteen grid with the aim of encircling their opponent's stones and controlling more territory. From this fundamental concept comes fiendish complexity. There are many more possible combinations on the Go board than atoms in the universe; in fact, there are many more possible combinations on the Go board than total atoms if every atom in the universe had an entire universe of atoms within it.[14] No database of human games was big enough, no set of human-generated rules complete enough, and no computer fast enough to handle anything close to all these possibilities.

Even more so than chess, Go is a game of intuition—the art of making decisions based on experience and pattern recognition rather than on logic and calculation. Psychologists have studied human intuition and found that, for some complex and high-pressure problems, experts can develop a feel for which course of action is best even if they are not aware that they are making a decision or consciously evaluating options. One oft-cited example is expert firefighters who develop an intuitive sense for when a burning building is about to collapse.[15] Such intuition, the capacity to decide without explicitly reasoning through a problem, is remarkable.

In Go, players must use their intuition to sort through the vast possibilities. They must balance important regional battles for control of smaller spaces with a broader strategic contest to dominate the board. Given the difficulties of calculation, top Go players often make moves because they "feel right."[16] The ability to make such moves garners great respect in China; Confucius commented on the game, and mastery of it was considered one of the four essential arts of Chinese scholars, dating back to the Tang dynasty.[17] Given the challenge of Go, the *New York Times* quoted computer scientists as saying that it would take a century or two of sustained progress even after Deep Blue's victory before computers would be capable of victory in the more complicated game.[18]

Across the Atlantic Ocean in the United Kingdom, another games prodigy was aware of the Deep Blue match and what it meant. Born to immigrant parents who were teachers and had once owned a toy shop, Demis Hassabis used chess tournament winnings to buy his first computer at the age of eight. By thirteen, he was a chess master and the world's second-highest rated player in his age bracket. By seventeen, he was the lead programmer on *Theme Park*, a well-regarded video game that helped pioneer the simulation genre. At the time of Kasparov's defeat in 1997, Hassabis was twenty years old and a star computer science student at Cambridge University. The cocreator of the World Wide Web, Tim Berners-Lee, would later describe Hassabis as one of the smartest human beings on the planet. He won the Pentamind competition—which crowns the best all-around board game player in the world—four years in a row, and five years out of six.[19] He was inspired by Deep Blue's victory and told a Japanese board game master that he would someday try to create a program capable of doing something similar for Go.[20]

In 1998, after graduation and a short stint working for the legendary video game designer Peter Molyneux, Hassabis cofounded Elixir Studios with David Silver, a friend from Cambridge whom Hassabis had once taught how to play Go.[21] Their company sought to produce pathbreaking video games featuring cutting-edge AI. Their two creations, *Republic* and *Evil Genius*, earned praise for their ambition but never achieved blockbuster commercial success. In 2005, the pair sold off the company's assets. Hassabis went to University College London for a PhD in neuroscience, while Silver went to the University of Alberta for a PhD in AI.[22]

Silver's PhD focused on a class of AI algorithms called reinforcement learning.[23] Like most advanced supervised and unsupervised learning algorithms, reinforcement learning algorithms often depend on neural networks. Unlike supervised learning, though, reinforcement learning does not rely on human-labeled training data; unlike Goodfellow's GANs, it does not involve two specialized neural networks in competition with one another. Instead, reinforcement learning algorithms center on the interaction between a computer program, called an agent, and its environment. In a given environment, the agent makes a decision that prompts a series of events. In the case of a virtual board game, for example, the agent could choose to play a particular move or initiate a sequence of moves. In a real-world environment, the agent controlling a robot could direct the machine to bring an object from one room to a person in another.

The environment then responds. For example, another player in the game could make a move or the person could take the object from the robot. The agent observes the state of the changed environment before it moves again. As the game ends or the person approves of the object offered by the robot, the agent eventually learns whether its decisions led to success or failure. It uses this information to improve its decisions in the future. Mathematically, the agent optimizes for choices that achieve the best possible results, learning which decisions and combinations of decisions lead to the desired outcomes. Over time, the agent refines its neural network, adjusting parameters so that it will, on future occasions, choose the appropriate course of action.[24]

Go's immense number of possible combinations posed a challenge for reinforcement learning algorithms. There were just too many options to calculate the optimal decisions using conventional methods. At any given moment, an agent would have to choose from hundreds of possible moves while also trying to determine how its opponent would respond when choosing among an equally large number of possibilities, and then what the agent's reply would be, and then its opponent's next move, and so on.

Even more challenging was that Go, in contrast to games like basketball or, to some degree, chess, doesn't have a clear scoreboard. At any given point in time, it is hard to calculate which Go player is winning; this, like so much else in Go, is a matter of intuition. Only at the end of the game would the agent find out if its choices had resulted in a win or a loss. It would then have to go back and figure out which moves along the way had helped and which had hurt. Unlike in many other sports and games, the impact of individual decisions in Go is often difficult to determine.

Silver focused much of his PhD research on designing an algorithm that could play Go at a master level. But the reinforcement learning agents he created were only able to sustain this high standard of play on a Go board that was nine squares by nine squares—one-quarter the total size of a traditional board and thus with many fewer possible moves and combinations of moves. While Silver's algorithm was a remarkable achievement that helped springboard him to a position as a lecturer at University College London, it nonetheless suggested that AI's true mastery of Go was still many years, if not decades, away.

In 2010, after completing his PhD and a stint as a visiting scientist at Harvard and MIT, Hassabis and two other friends founded an AI company called DeepMind. Silver was involved as a consultant from the beginning and joined the London-based firm full-time in 2013. In contrast to other AI startups that tended to be narrower in scope, Hassabis set an ambitious goal for DeepMind: "To solve intelligence, and then to use that in order to solve everything else."[25] It is a sentiment that captures what evangelists believe AI can offer to humanity. On the list of things Hassabis thought AI could help solve were "cancer, climate change, energy, genomics, macroeconomics, financial systems, physics." In each of these fields, he argued, "There's such an information overload that it's becoming difficult for even the smartest humans to master it in their lifetimes. . . . [Future AI] will automatically convert unstructured information into actionable knowledge. What we're working on is potentially a meta-solution to any problem."[26]

To achieve such a bold goal, DeepMind chose to focus largely on reinforcement learning, an unpopular choice by the research standards of the day. Hassabis said he was inspired in this choice by the human brain. Of all the types of algorithms, reinforcement learning seemed most similar to the dopamine-based mechanisms that humans employ and that Hassabis had studied as a neuroscience PhD student. If nature had found a use for that type of algorithm, so could DeepMind.[27]

It was unlikely that many traditional venture capitalists would invest in such an ambitious proposition. A startup that aimed to solve any problem in the world by mimicking the brain sounded closer to the stuff of science fiction or crackpot conspiracies than it did to a valid business model. DeepMind's cofounders decided to focus on pitching billionaires, reasoning that they would have the capital to fund outlandish ideas and the patience to wait a long time for results.

After some research, they found just such a billionaire. Peter Thiel was the cofounder of PayPal and an early investor in Facebook. He had also been a noted chess player when he was younger. Hassabis schemed a way to meet Thiel, eventually securing an invitation to a conference he was attending. As hundreds of people tried to pitch Thiel on their startups, Hassabis instead took advantage of a brief window to talk about chess. The brilliance of chess, Hassabis told Thiel, was that the knight and the bishop were in perfect strategic tension, each valued roughly the same but vastly different in their precisely balanced abilities. Thiel was so struck by the idea that he invited Hassabis back the next day to pitch

DeepMind. Eventually, Thiel, alongside other big names like Elon Musk, invested.[28]

With cash in hand in 2011, Hassabis and his team got to work. While AlexNet and its successors were racking up victories with their supervised learning approach and Goodfellow was using GANs to teach computers something akin to imagination, DeepMind began to popularize reinforcement learning with a series of surprising successes. In 2012, the company developed a system that was capable of playing many older video games better than even the best human players.[29] Remarkably—and in contrast to Deep Blue's extensive human instructions—DeepMind provided its reinforcement learning algorithm comparatively little information about each video game. The agent learned through trial and error, testing out different approaches to playing and observing which led to the best results. Early in its training, the agent fared terribly, but with each iteration the agent got better and better and better.

In 2014, Google bought DeepMind for reportedly 400 million British pounds, or more than $600 million. Google made clear that it didn't see its new purchase as an immediate profit center but rather as a long-term bet on the future of AI, and the company invested large sums of money to help that bet pay off.[30] With expanded resources, talent, and computing power, DeepMind was ready to try its hand at Go. David Silver took charge of the team that would develop the agent, called AlphaGo, and test just what reinforcement learning algorithms could do.

To begin, DeepMind built a supervised learning neural network. As training data, the company's engineers provided the network with thirty million positions—snapshots of the arrangement of the pieces on the board—derived from some of the world's top Go players' games. The network had thirteen layers of neurons that were together tasked with predicting which move a top human player would be likely to make in any given position. The network's guess would then be compared against the move that the human actually played. After four weeks of training, during which the system adjusted the strength of the connections between its neurons countless times, the supervised learning algorithm could guess the move selected by a human 57 percent of the time. This was impressive, beating the previous world record of 44 percent, but it was nowhere near good enough to defeat the world champion at Go, as it meant the algorithm could mimic a top human player only a little more than half the time.

DeepMind then took this neural network and pitted it against a copy of itself—a concept called self-play. Whereas GANs feature two different neural networks specialized for distinct tasks, DeepMind had the two very similar versions of AlphaGo battle it out on opposite sides of the virtual Go board. Using the trial-and-error processes of a reinforcement learning algorithm and many games—each played very rapidly by the two versions—AlphaGo began to learn which moves were more likely to lead to success at the end of the game. "AlphaGo learned to discover new strategies for itself, by playing millions of games between its neural networks, against themselves, and gradually improving," Silver said.[31] After this training, the system was better at Go than most of the state-of-the-art computer programs in the world—but not yet better than the top humans.

By this point, DeepMind had solved a fundamental problem: its system could, in any given Go position, eliminate most of the possible moves and focus just on the ones that seemed best. This was a significant advance, as at any given time there are usually around two hundred possible moves in Go (as compared to twenty or fewer in chess). By focusing on just a few, and then just a few of its opponent's possible responses, AlphaGo was mimicking the intuitions that top humans deploy when playing Go.

Another problem loomed: Go games are long, sometimes taking several hundred moves to complete. Even if AlphaGo could reduce the complexity at each step, calculating all the way to the end of the game would become exponentially difficult, as each move added on more and more possible combinations. AlphaGo needed a way to develop intuition not just for determining the best moves at any point but also for estimating its chances of winning the game from any given position. This kind of automated intuitive evaluation of a position was thought to be impossible.

Hassabis, Silver, and the team at DeepMind proved the conventional wisdom wrong once again. They decided that they needed to build a second neural network within each version of AlphaGo that focused only on this task of evaluation. Rather than attempt to predict the best move for a given position, it would instead give its best guess about which side was favored to win if both sides played well. Crucially, it would do so with something akin to intuition—examining the position for its general patterns and their similarity to other positions it had seen—rather than by calculating through brute force every single move and countermove until the game ended. The network played thirty million virtual games against a copy of itself over the course of a week, using the refined trial-and-error process of reinforcement learning to get better at its task.

The combination of the two trained neural networks was powerful. In essence, they provided AlphaGo with a kind of holistic automated intuition; similar to how firefighters learned to intuit a burning building's structural integrity from experience, AlphaGo learned to translate its tens of millions of game experiences into Go insights. In any given Go position, AlphaGo could use its automated powers of intuition in two ways. It deployed the first neural network to focus on just a few possible good moves, what its opponent's response would be, what it would do in response, and so on. To avoid being overwhelmed with possibilities, though, it eventually turned to its second neural network, the one that evaluated its chances of winning in any given position without having to calculate all the way through to the end. Using the two networks together, AlphaGo could test out a bunch of possible moves in simulation, see what kinds of positions these moves would likely produce, and then estimate its chances of success in those imagined positions.[32]

With the system designed and trained, one question remained: Could AlphaGo beat the best players on Earth?

ALGORITHMIC VICTORY

To answer this question, DeepMind first arranged for AlphaGo to play Fan Hui, the three-time European Go champion. In October 2015, AlphaGo prevailed decisively, winning five games and losing none. It was the first time that a professional Go player had lost a game to an algorithm on a full-size board. DeepMind triumphantly announced the news in a major paper, published in January 2016 in the leading scientific journal *Nature*.[33]

But some argued that Fan Hui wasn't a stiff enough test for AlphaGo. Other humans were better players, and perhaps the best of them all was Lee Sedol, the winner of eighteen international titles. Lee agreed to play

AlphaGo, and the five-game match took place in March 2016 at the Four Seasons Hotel in Seoul. In South Korea, the games were broadcast everywhere, even on giant screens in the streets. Tens of millions watched it all over the world, with sixty million viewers in China alone. The audience included both die-hard Go fans and many viewers who may have known almost nothing of the ancient game but wanted to witness the newest battle royale between humans and machines.

Lee did not shy away from the stakes. In a pre-match interview, he said, "I'm going to do my best to protect human intelligence." He was confident he could deliver. "I don't want to be too arrogant, but I don't think that it will be a very close match," he said as he predicted either a 5–0 or 4–1 victory for himself.[34] Just before the first game started, he went silent and closed his eyes, as if to both clear his mind and focus on the scale of the task he was about to undertake.[35] Fan Hui would later say that Lee knew this game was different; usually he played for himself, but this time he was playing for all of humanity.[36]

AlphaGo won the first game with ease, shocking Lee and almost all observers. Toward the end of the game, Lee's wife and daughter looked forlorn in the audience; with her hands covering her face, Lee's daughter seemed to be crying or praying.[37] Lee's defeat marked the first time a top Go world champion had lost to a computer. It was a symbol of just how quickly and unexpectedly algorithmic progress had accelerated. Algorithms for playing Go had progressed from the level of human amateurs to defeating the world champion in seven years; similar progress in chess had taken thirty years.[38] Not even two years before the AlphaGo match, experts had predicted that the first victory for an algorithm against someone like Lee was at least a decade away.[39]

Game two was even more surprising. It showed that machines could not just mimic but surpass human intuition on the Go board. At a critical moment, move 37, AlphaGo made a bold and surprising choice: it launched an unexpected attack on Lee's right flank. Commentators, including some of the best Go players in the world, were confused. The Korean broadcasters of the match called it an unthinkable move, and the English-language expert said, "I thought it was a mistake." Lee, struck by what AlphaGo had done, thought for more than twelve minutes before responding.[40]

Perhaps the person who most immediately appreciated the move was Fan Hui, the very man an earlier version of AlphaGo had vanquished a few months before. He said he was initially stunned by the algorithm's attack, but after ten seconds started to guess at its sophistication and its power. Even he could not fully comprehend it, and later offered this explanation: "It's not a human move. I've never seen a human play this move. . . . So beautiful."[41]

At some level, AlphaGo recognized that its move 37 was not a human move. Its first neural network, the one trained to forecast what moves top Go players would make in key situations, could calculate as much. That network guessed that the odds of a top human choosing the same bold strike against Lee's right flank were 1 in 10,000. And yet, guided by its own tens of millions of games against itself and the refined insight those games provided its other neural network designed to evaluate positions intuitively, AlphaGo played the move anyway. The machine had grasped the limitations of human capabilities—and then exceeded them. Two and a half hours later, long after it was clear the game was lost, Lee resigned. Move 37 was decisive. One commentator said of AlphaGo's powerful attack, "Go is like geopolitics. Something small that happens here will have ripple effects hours down the road in a different part of the board."[42]

The postgame mood was somber. Lee said, "Yesterday I was surprised. Today I am speechless. Very clear loss on my part. From the very beginning, there was not a moment where I thought I was leading."[43] On the day off after the game, he went over the match with four professional Go players, trying to understand what had happened and how things had gone wrong for humanity.[44] Lee needed to turn it around quickly, since he now had to win three games in a row.

In game three, Lee changed his tactics, trying to be more aggressive against the AI and forgoing some of his usual patient style. He seemed to be uncomfortable the whole time, biting his nails and touching his face. A commentator noted he was "fighting a lonely fight."[45] None of it mattered to AlphaGo, which plowed ahead with ruthless efficiency and won decisively, clinching the match.

Afterward, Lee apologized "for being so powerless."[46] Hassabis said that he felt ambivalent; as a top game player, he knew how terrible it was to lose. Another DeepMind employee said, "I couldn't celebrate. It was fantastic that we had won, but there was such a big part of me that saw [Lee] trying so hard and being so disappointed."[47] A deep sense of melancholy pervaded the room. Perhaps for this reason, Hassabis announced before the fourth game that he was rooting against the algorithm he had helped create and for Lee.

In game four, it at first looked like Lee would need all the help he could get. AlphaGo began demolishing Lee's position with the same kind of brutal and methodical ruthlessness that it had deployed over the previous three games. On move 78, with no doubt that he was in trouble, Lee took more than thirty minutes to think. When he finished, he placed a stone near the middle of the board. It was such a surprising decision that it was later called "a divine move" among Go players, since it seemed no human could have thought of such a bold and brilliant response.

AlphaGo was stunned as well. The algorithm had calculated that the odds of a top human choosing the move Lee played were 1 in 10,000. AlphaGo butchered its response and, after trying and failing to continue the fight, eventually resigned. Lee had won. He said, "The victory is so valuable that I would not exchange it for anything in the world."[48] In the battle against the machine, there was hope for humanity.

To those looking for poetic beauty, here it was. The symmetry of move 37 in one game and move 78 in the other showcased the power of both humans and machines, and hinted at how they might work together. Each move was 1 in 10,000. The machine had found the former move despite calculating with near certainty that no human would; the human had found the latter move despite the machine estimating with nearly the same certainty that he would not. *Wired* magazine wrote, in an ode to the future partnership of algorithms and humans, "The machine that defeated [Lee] had also helped him find the way."[49] Lee said that AlphaGo taught him that "moves that we thought were creative were actually conventional," though he looked somewhat melancholy as he said it.[50]

But there is a very different story to be told, one that emerges only with the benefit of hindsight. It is a tale of the seemingly inexorable march of AI progress, of extremely rapid growth in capabilities showing itself in the world of algorithms once more, of the new fire raging ever onward. It begins with the simple fact that whatever insight or fortune Lee had discovered, it did not last. In game five, AlphaGo won again.[51]

The algorithm would only get better from there. Several months after the match with Lee, in December 2016, DeepMind quietly deployed a new version of AlphaGo into top-level online play under the pseudonyms Magister and Master. In the span of just over a week, this version of AlphaGo played sixty games against some of the best players in the world, including a string of national champions and world champion runners-up. It won every single one. Even more impressively, it beat the then-top-ranked player in the world, Ke Jie, three times. Near the end of this incredible run of victories, Hassabis announced that DeepMind was in fact behind the pseudonyms.[52]

This revelation did not come as a great surprise to the Go community, given the unconventional and unsparing tactics that Magister and Master had employed in their online matches. Ke himself studied the matches and remarked that the games had shown him just how far ahead algorithms now were in playing Go. "After humanity spent thousands of years improving our tactics, computers tell us that humans are completely wrong. . . . I would go as far as to say not a single human has touched the edge of the truth of Go," he said.[53] A few months later, in April 2017, he agreed to play against the algorithm in person in Wuzhen, China. DeepMind offered him a $1.5 million prize if he won.

But this time there would be no poetic beauty, no machines showing the way for humans to leapfrog ahead, and no real reason to doubt algorithmic supremacy. The new version of AlphaGo was even stronger than the one that had defeated Lee. It demolished Ke, decisively winning three games and losing zero; at one point, Ke wiped tears from his eyes as it became clear he would lose.[54] Another world champion had fallen. After the match, DeepMind announced that it was retiring AlphaGo from competitive play.[55] The company didn't explicitly say it, but the reason was clear: no one left stood a chance.

Or, more precisely, no human stood a chance. Unbeknownst to almost anyone, the version of AlphaGo that demolished Ke in China was not actually the best system in the world. There was another one that was even better, still hidden away in DeepMind's research lab. It was almost ready to burst into view.

ALGORITHMIC INDEPENDENCE

One way of looking at AlphaGo held that human knowledge—in the form of the thirty million positions from top players' games—served as its main foundation. Another interpretation saw human knowledge as its main hindrance. DeepMind took the latter view.

As AlphaGo was eviscerating all human competition, the company began work on a successor called AlphaGo Zero, another reinforcement learning system. DeepMind provided this algorithm with the rules of Go and nothing more. Rather than feeding it the same thirty million positions of top human play that the company had given AlphaGo in its early stages of development, DeepMind's engineers gave AlphaGo Zero only the capacity to play against itself time and time again, learning a little bit from each game. Whereas AlexNet learned everything that it would ever know from human-labeled data, AlphaGo Zero would learn everything that it would ever know about Go strategy from the self-generated data of its own games.

And there were a lot of games. AlphaGo Zero played 3.9 million matches against itself, starting out with essentially random play and getting better and better over time. In this way, it mimicked a human player, learning the game over a lifetime; intermediate versions of AlphaGo Zero showed that it incrementally added new tactics, concepts, and strategies to its repertoire. The difference was that AlphaGo Zero learned from playing itself very rapidly, discovering the game as it did, whereas humans often learn slowly from other humans.

What AlphaGo Zero lacked in a foundation of data, it more than made up for with algorithmic power. DeepMind made several improvements that enabled the algorithm to develop an even stronger automated form of the intuition required to play Go. First, the company better enabled the code to adjust its approach to Go based on how its strategies were faring against copies of itself in training. This was essential, as AlphaGo Zero would not need human strategies if it could quickly update itself to learn which decisions were more likely to result in wins. This capacity to get just a little bit better with each iteration compounded over 3.9 million games, more than a human could play in many lifetimes, to help the algorithm become the best player in the world.

Second, DeepMind made an unusual, though effective, decision: it combined the two neural networks that AlphaGo used into one, and then divided the output layer. After all of the network's layers—the body—the network essentially sprouted two heads. One head was optimized to predict the probability of a player making certain moves, and the other was optimized to predict the probability of winning the game. Though these calculations were different, they relied on a lot of the same insights about the game of Go and so benefited from sharing the same core network layers. In this sense, DeepMind designed the algorithm to be more flexible, in keeping with its mission of trying to build general intelligence that could perform many tasks. It was also perhaps more humanlike, as the human brain is adept at transferring learned skills between related activities; when a tennis player learns to swing a baseball bat, she doesn't need to start completely from scratch.

Third, DeepMind empowered AlphaGo Zero with some of the latest neural network techniques. These improvements included some tricks that had not yet been invented when the company started work on the original AlphaGo. Chief among these was an architecture known as a residual neural network, which adjusted how the data cascaded through the system, sometimes allowing it to skip over layers in an effort to improve training. The new design was a foundational part of Microsoft Research Asia's success in the 2015 ImageNet competition.

With these algorithmic improvements, the final version of AlphaGo Zero got very good very fast. After just three hours of playing against itself and training its neural network, AlphaGo Zero was better than most human beginners. Within a day, it had learned many of the more advanced Go strategies. Within three days, it was better than the version of AlphaGo that had defeated Lee Sedol. After twenty-one days, it was better than the improved version of AlphaGo that had beaten Ke Jie. AlphaGo Zero had learned more about Go from itself in three weeks than all of humanity had been able to learn across millennia. Such was the power of exponential growth.[56]

But for the evangelists at DeepMind, the mission was not to win at board games; it was to solve intelligence and use that to solve problems facing humanity. As the next step, DeepMind aimed to create an algorithm that was even more flexible than AlphaGo Zero, one that showed greater progress toward generalizability. The new algorithm would learn to beat any human or algorithm at not just one board game but three: Go, chess, and a Japanese version of chess called shogi. DeepMind wanted its more general system to triumph even over specialized opponents that focused on just one game. As with AlphaGo Zero, the company would provide its creation with only the basic game rules and nothing more.

In December 2017, DeepMind introduced AlphaZero. It used many algorithmic principles similar to AlphaGo Zero, but made allowances for the ways in which chess and shogi differed from Go. Whereas Go always has a winner and a loser, the other two games feature draws; in addition, the other two games have position-dependent quirks, like the fact that pawns can initially move two spaces in chess.

AlphaZero adapted to these nuances better than any other algorithm on Earth. It decisively defeated Stockfish, an expert system-based chess program that had been in development for more than a decade and was widely considered to be the best in the world. Against Stockfish, AlphaZero won 155 games, lost 6, and drew the rest. It deployed an efficient algorithm to win, in contrast to Stockfish's brute force: AlphaZero focused its analysis on the most promising 60,000 positions per second while Stockfish, lacking such a refined approach, analyzed 60 million positions per second, the vast majority of which were irrelevant.[57] At shogi, AlphaZero defeated the best competitor algorithm 91 percent of the time. In Go, it defeated its powerful predecessor, AlphaGo Zero, in 61 percent of their matchups. DeepMind did not bother testing AlphaZero against humans, since everyone knew who would win, and it wouldn't even be close.[58]

It was not just the fact that AlphaZero was winning that was striking, though. It was how. Whereas previous expert systems had seemed mechanistic and rote in their success, AlphaZero was dynamic and flexible, even beautiful and alien. The Danish chess grandmaster Peter Heine Nielsen said, "I've always wondered how it would be if a superior species landed on Earth and showed us how they played chess. Now I know."[59] Another chess grandmaster, Matthew Sadler, said, "It's like discovering the secret notebooks of some great player from the past."[60] The machine was capable of teaching even the best humans on Earth how to play better.

AlphaZero's rate of learning was faster than AlphaGo Zero's, even though it had to handle three games. With nine hours of training, it mastered chess, playing a game every three seconds.[61] With twelve hours of training, it mastered shogi. With thirteen days of training, it mastered Go, the most complicated of them all.[62] Once again, the strategic insights that humans had gleaned over millennia were rendered impotent next to what an algorithm could learn in hours, days, or weeks. All that players had ever known about these games was—to this rare machine of intuition and calculation—simply irrelevant, if not wrong.

The achievements of AlphaZero should prompt modesty about humanity's comparative ignorance, but they should also inspire wonder at what humanity's algorithmic creations can do.[63] For Hassabis, Silver, and others, the sequence of research projects that began with AlphaGo represented further proof of accelerating algorithmic power, and suggested that capable enough algorithms might someday attain general intelligence. The big bet that DeepMind had taken on reinforcement learning had clearly paid off.

If the ImageNet competition had put AI on the national security radar screen, AlphaGo and its successors triggered a red alert. An easy narrative started to take shape, especially after the defeats of Ke Jie and others: a Western company's algorithm had beaten the world's best players at a game that had originated in China. Given that Go had such obvious, though abstract, connections to military strategy, it was tempting to assume that Western dominance on the game board presaged a similar strength on the real-world battlefields of the future, and perhaps even in geopolitics more generally. While this narrative lacked much technical rigor, it seems to have gained popularity in China.[64]

One analogy quickly dominated the conversation: *Sputnik*, the space satellite that the Soviet Union launched in October 1957. Tactically, it offered almost no benefit to the Soviets. The satellite was a sphere less than two feet in diameter. It had four radios that broadcast its location at regular intervals and batteries that lasted for three weeks, after which the satellite went silent. A few months later, it burned up on reentry to Earth's atmosphere.

Strategically, however, *Sputnik* mattered quite a lot—for the United States. *Sputnik*'s success made for a striking juxtaposition with the spectacular failure of the United States' first satellite launch in December 1957. As *Sputnik* orbited, alarm bells rang throughout Washington. Congress began weeks of hearings—many led by Lyndon Johnson, the famed senator and future president—on the supposed "missile gap" between the two superpowers. Within a year, policymakers founded NASA and the Advanced Research Projects Agency. Within four years, President John F. Kennedy promised that the United States would land a man on the Moon and return him safely to Earth. Within twelve years, *Apollo 11* succeeded in that feat.[65]

AlphaGo, the argument went, had done for China what *Sputnik* did for the United States. It established AI as a new terrain of geopolitical competition, one with obvious implications for war and peace. It accelerated Chinese government investment in AI to the tune of billions of dollars each year and drove more students into the relevant fields in math and science.[66] It meant that, once more, nations would compete to see who could push the boundaries of technology the furthest.[67] But as the warriors sounded the geopolitical klaxons, the evangelists at DeepMind charted a different course for AI.

ALGORITHMIC VALUE

Hassabis had always wanted DeepMind to do more than just play games. He thought AI should benefit all of humanity, unlocking the universe's secrets and solving thorny problems. He believed that scientific discovery was central to the company's mission: when designing the firm's new headquarters, he placed a reference to DNA, in the form of a helix-shaped staircase, at its center. Success at Go was impressive, he thought, but it was only a milestone on AI's march to something greater.

To advance this mission, DeepMind's leadership gathered a group of employees for a two-day hackathon. As AlphaGo was crushing all opposition in the spring of 2016, the assembled scientists and researchers considered at an informal brainstorming session which project to pursue next. These hackathon participants wanted to find a problem that would not just showcase DeepMind's capacity for algorithmic innovation but also prove that such innovation could have a positive impact on the real world.

A group of three employees came up with an intriguing idea. They would solve the protein folding problem, which had vexed scientists for decades.[68] Only one ten-billionth of a meter in size, proteins are behind nearly everything that happens in living bodies: they maintain the structure of cells, kill disease, digest food, transport oxygen, move muscles, and much more.[69] When a cell first produces a protein from its genetic code, the protein is a thin strand or coil of amino acids. Within milliseconds, the protein arranges itself (or folds) into a specific 3D shape that enables it to perform its function within the body. Predicting the structure of a protein from a known amino acid sequence is difficult. One leading molecular biologist estimated that there were so many possible folds for any one protein that it would take more time than the age of the universe to list them all. Using advanced X-ray machines and microscopes, a PhD student might dedicate an entire doctorate to discovering how just one protein folds.[70]

An imperfect analogy can provide some insight into the difficulty of predicting a protein's 3D structure from its sequence of amino acids. The amino acids in a protein chain are akin to pearls on a string, with each of the twenty possible amino acids represented by a different color of pearl. When the string lies on a table, it is easy to look at the colors and see the sequence of the pearls. It is much harder to imagine what the shape of the string of pearls would look like if the pearls, buoyed by a range of nongravitational forces causing some pearls to be attracted to or repelled by others, were to leap off the table and into a 3D arrangement with one another. The distance between each pearl and the overall shape of the 3D whole would depend on this array of forces. They would be easier to observe than to predict.[71]

Developing a better model for protein folding would have a tremendous impact on medicine. Scientists believe that diseases like Parkinson's and Alzheimer's result in part from proteins that do not fold properly; a better understanding of this process might help treat those diseases.[72] Even after painstaking work manually studying proteins, scientists only knew the structure of about half of the proteins in the human body—a huge gap in knowledge that obscures key health interventions and insights.[73] Improved models for protein folding could also help with drug discovery, devising new medications to treat a range of conditions and reducing the side effects of treatment. Similar insights could enable biomedical engineers to more easily synthesize proteins. One scientist suggested that a more detailed understanding of proteins could enable synthetic creations that digest plastic and reduce waste.

Despite the difficulty of modeling protein folding, DeepMind's scientists had reason for optimism when they considered how AI might help. Unlike some of the other hard problems that the employees considered, the study of protein folding offered large and well-organized data sets. Researchers had collected the sequence and observed the 3D structure of more than 150,000 proteins, each observation the result of extensive study. They assembled this information in the Protein Data Bank—the perfect training data for a machine learning system.

Hassabis greenlit the project, which acquired the name AlphaFold, and liked that making progress on understanding protein folding would have a broader impact on the world. "We try and find root node problems—problems where if you solve them it would open up whole avenues of new fields for us and other people to research," he said later.[74] DeepMind had officially joined the world of science.

The company now needed to show what it could do. In protein folding, the leading competition is known as the Critical Assessment of protein Structure Prediction, or CASP. It is held once every two years and draws more than one hundred participants from research labs all over the world. The organizers privately determine the structure of proteins that have not previously been studied in public, then task competitors with, among other things, modeling the various biochemical forces and predicting the 3D structure of the proteins from their amino acid sequence. The winner is the one who comes closest to guessing nature's shape for each protein.

With everyone working from the same data set in the Protein Data Bank, what matters in the competition is algorithmic innovation. DeepMind trained a neural network that looked at the amino acids encoded in a protein's amino acid sequence and made a prediction about how far away from each other they would be when the protein assumed its 3D shape. A second neural network evaluated the range of possibilities from the first neural network and determined which were likely to be closest to the right answer.

AlphaFold then used two methods to refine this initial guess. One approach searched the Protein Data Bank for proteins that were similar to the one being modeled and used yet another neural network to generate different possible fragments from the proteins it found. The second method iteratively refined the proposed structure of the protein, similar to the way in which supervised learning networks adjust their parameters to improve their performance. After these refinements, AlphaFold came up with its best guess of how the protein would fold. The technique seemed to work in practice, but the expert competitors at CASP in December 2018 would offer the real test.

Like AlphaGo before it, AlphaFold dominated its competitors. When tasked with predicting shapes from sequences, AlphaFold came up with the best structure for twenty-five of forty-three proteins; the runner-up entry managed the best score only three times. Mohammed AlQuraishi, a leading protein folding expert, estimated that DeepMind had made twice as much progress as usually occurred between CASP competitions. Put differently, DeepMind's algorithm had achieved in only two years the technological advancements that would have been expected to take close to four years. It was a giant leap forward.[75]

AlQuraishi noted the pivotal nature of DeepMind's triumph. "We have a new, world class research team in the field, competitive with the very best existing teams," he wrote. An AI company with no scientific pedigree and a research group of around ten people had used its algorithms to outperform multibillion-dollar pharmaceutical firms and university labs staffed with the top biologists in the world. It was as if a biotechnology giant had in a short period of time become the best in the world at computer speech recognition or building self-driving cars.[76]

DeepMind's progress only gathered speed from there. In November 2020, less than two years later, the company announced that it had solved the protein folding problem—a monumental scientific achievement. It revealed AlphaFold 2, a system that was able to determine with a high degree of accuracy the 3D structure of any protein when given its amino acid sequence. AlphaFold 2 could handle in a few days the kind of scientific inquiry that once required hundreds of thousands of dollars and years of effort. No other human or automated computer system was close; in an ironic twist, AlphaFold 2 helped human judges of the 2020 CASP find the structure of four proteins that they had not yet worked out.[77]

The scientific community was stunned. Even after DeepMind's remarkable 2018 success, AlQuraishi had thought that solving the protein folding problem to this degree was still more than a decade away.[78] "Never in my life had I expected to see a scientific advance so rapid," he wrote. "The improvement is nothing short of staggering."[79] John Moult, one of the creators of CASP, said, "I always hoped I would live to see this day, but it wasn't always obvious that I was going to make it."[80] Andrei Lupas, a leading biologist, used AlphaFold to find the structure of a protein that had eluded his research team for a decade. Afterward, he said, "This will change medicine. It will change research. It will change bioengineering. It will change everything."[81] In 2021, DeepMind began to deliver on this promise, releasing the protein structure for more than 350,000 proteins—including nearly every protein in the human body—with plans to release the structure for another 130 million.

DeepMind's success bucked a broader worrying trend: science seems to be slowing down.[82] On key research questions outside of AI, the amount of effort required for new research is increasing even as the overall rate of advance slows. One survey of Nobel Prize–winning research contends that breakthroughs from decades ago are more impactful than comparable advances today. Another survey of top scientific researchers from the last forty years found that their best discoveries often came later—at an average age of forty-seven rather than thirty-seven more than a century ago—in part because they have to learn so much about their field and subfield just as a foundation for new insights.[83]

Modern science also requires far more people and far more complex collaboration. A century ago, the paper that announced the discovery of the atomic nucleus had a single author; in contrast, a notable particle physics paper in 2015 had more than 5,000 listed authors.[84] On average, the size of leading scientific research teams has nearly quadrupled in the last hundred years.[85] In addition, some observers worry that the rate at which scientific breakthroughs improve modern life is decreasing, especially outside of information technology.[86] When DeepMind won CASP, the traditional return on research and development efforts in fields like drug discovery and biopharmaceuticals was substantially lower than in past decades.[87]

To AI evangelists at DeepMind and elsewhere, machine learning offers a rare opportunity to reverse this trend. Science can benefit from the increasing capacity of machines to process vast stores of information and approach problems from a variety of creative directions, some of which are new to even those humans who have studied the same problems for decades.[88] For example, one DeepMind project in progress uses machine learning to solve equations in quantum physics and chemistry that are not solvable by current approaches; such an effort, if successful, would unlock fundamental scientific insights and have immediate practical utility.[89] Other AI research labs have made progress with applying machine learning to key technological challenges, such as improving the utility of renewable energy systems, gaining greater insight into earthquakes and floods, strengthening climate data collection, and increasing battery efficiency.[90]

As ever, while the warriors try to translate the advances in machine learning to the geopolitical arena, evangelists like Hassabis have their eyes on a different prize. "Our mission should be one of the most fascinating journeys in science," he said. Even before AlphaFold's success in the 2018 CASP, Hassabis thought the journey would reach great destinations. "I think what we're going to see over the next 10 years is some really huge, what I would call Nobel Prize–winning breakthroughs, in some of these areas," he said in October 2018.[91]

From AlexNet to AlphaFold, both the evangelists and warriors mostly saw reason to believe that AI fit their worldview. ImageNet and GANs showed the power of data-intensive machine learning, kickstarting a technological revolution but also fostering the automation of intelligence analysis and the creation of fake but convincing videos. AlphaGo and its successors demonstrated the strengths of algorithms, proving that machines were capable of outperforming humans at strategic tasks and prompting massive government investment to wield algorithms in the service of national security. AlphaFold showed the enormous potential of AI for medicine; the potential use of the protein folding technology for malicious purposes are as yet uncertain.

Worldviews aside, a look across the decade of machine learning algorithms shows just how far they have progressed. In 2012, the idea that
neural networks could achieve image recognition seemed far-fetched to
most; eight years later, the networks were capable of unlocking some of
science's fundamental mysteries. Alongside the tremendous growth in
algorithmic capability, however, was an exponential growth in algorithmic efficiency. In essence, the algorithms were getting both leaner and
more capable, needing fewer calculations to get the same performance. As
a result, a study in 2020 found that it took forty-four times less computing power to get an algorithm to reach the same level of performance as

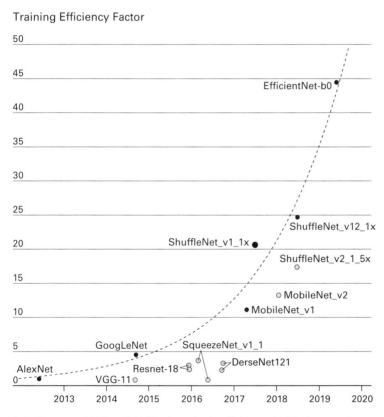

2.1 Between 2012 and 2019, algorithmic efficiency doubled every sixteen months,
with researchers at OpenAI measuring a forty-four-fold decrease in the amount of computing power needed to train a machine learning image recognition system to the accuracy level of AlexNet on ImageNet.

AlexNet had achieved on image recognition in 2012, with progress poised to continue.[92]

This increased efficiency dovetailed with another trend: computing power was becoming more accessible. After data and algorithms, computing power was the third spark of the new fire, enabling the continued improvement of machine learning systems. Even as DeepMind and others pushed algorithms to new heights of capability and efficiency, it was time for that last spark to become a flame.

3

COMPUTE

Every three seconds, a human fingernail grows about three nanometers. It is not much, but it is the scale at which some of machine learning's most complicated and important technologies—transistors—are manufactured.

Transistors make computer chips function, and computer chips perform the calculations that make machine learning algorithms work. Computer processors use these algorithms to calculate how to adjust the neural network as it learns from data, extracting insights and encoding them in the connections between neurons. As neural networks get larger and data sets swell, training machine learning systems requires more calculations; without the brains provided by computers, machines do not learn. Computing power—more commonly known as "compute" by AI researchers—is thus fundamental to modern machine learning.

The importance of compute can be disconcerting and even demoralizing to some algorithm developers. Richard Sutton, a pioneer of machine learning, called compute's central role in AI "the bitter lesson."[1] For as much as AI scientists want to invent intelligence in their labs—crafting thinking machines in humanity's own image—history has repeatedly shown that what matters most is devising a generally capable algorithm and providing it with more compute. In this view, more and better chips, not algorithm developers' ingenuity, have driven progress toward more powerful AI.

But making more powerful chips depends on making better transistors. A transistor is either on or off, depending on whether or not electric current is flowing through it. At a hardware level, transistors contain semiconductors, materials that can both carry and insulate electricity; semiconductor is also the name given to the final manufactured computer chip product. At a software level, binary logic provides a mechanism for the device to perform calculations using only two possible states (1 for on and 0 for off).

Smaller transistors are better. As transistors decrease in size, they can switch between on and off states more quickly, increasing the number of calculations the chip can perform per second. With smaller transistors, engineers can also fit a greater number of them on each computer chip, allowing more calculations to be performed. In 1965, the future cofounder of Intel, Gordon Moore, realized that engineers were capable of doubling the number of transistors in a processor every two years or so, leading to a rapid increase in processing speed. This exponential growth in transistor density became known as Moore's Law.[2]

Manufacturing ever-smaller transistors is hard. It begins with design. In the earliest days, when transistors were big and few in number, engineers could draw chip designs by hand. Later, as chips grew to include billions of tiny components, design became a world unto itself, one requiring highly specialized software and expertise. Even with sophisticated tools, designing a new chip often takes months or years and costs hundreds of millions of dollars.[3]

After design comes fabrication. Fabrication facilities, or fabs, contain some of the most cutting-edge hardware in the world and feature advanced techniques in electrical engineering. Fabrication once happened on a scale of micrometers, but today it is on the scale of nanometers, just a few dozen atoms in length. It must take place in a sterile room, since a single speck of dust can ruin the process. It is so intricate and requires such expensive equipment that Moore's Second Law addresses it: as the chips get more advanced and more complex over time, the cost of building a factory to make them doubles every four years. This doubling exponentially increases the expense and technical challenges over time— and centralizes power within the few companies—and countries—that can handle those challenges.[4]

Within each fabrication facility, the process of fabricating chips is often more than a thousand steps long. It begins with circular wafers of silicon, an element commonly found in sand, that are a foot in diameter. The silicon is mixed with other chemicals to give it certain kinds of electrical conductivity. An advanced piece of equipment called a photolithography machine draws the design of the chip, which then gets imprinted onto the silicon.[5] These machines are finicky, often costing more than $150 million and requiring forty shipping containers to move.[6] Once the machines imprint the design, other components place metal on top of the silicon and etch electrical circuits into the metal.

The chip manufacturing process is tricky and technical; it also matters enormously for geopolitics. The semiconductor industry produces around one trillion chips per year, from the basic ones inside cheap electronics to the highly specialized ones in data centers that power machine learning systems.[7] As a result, the ability to manufacture or acquire vital computer hardware is central to national power. The nations that have mastered this intricate industry are democracies, and they enjoy strategic opportunities as a result. It is an advantage over autocracies that has held for decades, but it has taken on far greater importance in the age of AI. To see why semiconductors and the advantage they confer matter, it is first necessary to understand how a revolution in AI computing began.

A QUIET REVOLUTION

Andrew Ng, a noted machine learning researcher, was worth $1.5 billion to the Chinese technology giant Baidu—or so the markets said. On the day he left the company in 2017 to launch his own ventures, the firm's value dropped by that staggering amount.[8] The plummeting share price was a sign of how rare top technical leadership is in machine learning, how important cutting-edge technology is to modern business, what a titan Ng was in the field, and how far he had come to be worth so much to any company that hired him.

Ng's machine learning story began two decades before. Born in London in 1976 to parents who had immigrated from Hong Kong, he earned his bachelor's degree at the top of his class at Carnegie Mellon University, where he triple-majored in computer science, statistics, and economics.

After a master's degree at MIT, Ng earned his doctorate in machine learn-ing from the University of California, Berkeley, quickly producing a thesis that became a widely cited work of reinforcement learning.[9] After that, he moved to Stanford as a professor, where he started to gain wide recogni-tion in the field; among his students was Ian Goodfellow.[10]

Ng faced a persistent issue in his research: individual computer proces-sors weren't fast enough to train the neural networks that he and many of his contemporaries were dreaming up. To make machines learn what he wanted them to learn, he needed significantly more computing power. One way to do that was to put more processors on the job. He didn't want an algorithm to run on just one chip, or ten, or even a hundred; none of those would provide enough firepower. He wanted thousands of proces-sors training his neural networks.[11]

This idea, called parallelization, had been around for a long time in traditional computer programming.[12] Sequential processing requires the steps of an algorithm to be performed one after the other. Parallel proc-essing lets many steps be performed at the same time, cutting the time needed to perform calculations by increasing the number of machines at work. Parallelization worked for problems that could mathematically be subdivided into parts that would then be calculated independently. For example, it is possible to subdivide 1 + 2 + 3 + 4 into 1 + 2 and 3 + 4, instruct two processors to calculate one subproblem each, and then add the results.[13] Ng recognized that training neural networks could utilize parallelization because the training data could be subdivided into batches and processed in groups. He could have thousands of processors working on pieces of the same problem mostly on their own, each training the same neural network and incrementally making it more capable. While others debated the merits of this idea, Ng set off to make it happen at a massive scale.

Ng needed a lot of money and a lot of computers. Google had both. In 2011, Ng began a sabbatical at the company's X lab in California, which housed some of the company's most secret and ambitious projects. At X, he embarked on a bold effort that would show what many computer chips working in tandem could do. His team assembled 16,000 central processing units, or CPUs, and linked them together into a massive clus-ter of computing power. The total cost was over $1 million.[14]

To complete this project, Ng teamed up with Jeff Dean, one of the most famous engineers in Google history and one of the company's AI pioneers. When he was in high school in the 1980s, Dean coded a method for optimizing statistical analysis for the Centers for Disease Control.[15] In 1999, he joined Google as employee twenty-five, and he had a hand in nearly every notable piece of software the company produced for the next twenty years. He was so renowned that Googlers traded satiric "Jeff Dean Facts" with one another. These included "Jeff Dean's PIN is the last four digits of pi" and "When Alexander Graham Bell invented the telephone, he saw a missed call from Jeff Dean."[16] In 2012, Dean could have worked on virtually any computer project he wanted, but he chose to focus on AI with Ng.

Ng and Dean lifted still frames from ten million YouTube videos and designed an algorithm that would look for objects that appeared repeatedly throughout the images. This task required a gigantic amount of compute. The neural network Ng and his team designed featured more than one billion connections between its neurons, which it used to extract patterns from the massive data set. After spending several days processing the images, the system began to recognize the objects within them. Given that the system looked at YouTube videos, the first object it recognized was predictable: a cat. "We never told it during the training, 'This is a cat,'" Dean said. "It basically invented the concept of a cat."[17]

Ng wasn't done. Even before his experiment at Google, Ng had brainstormed ideas to grow the compute used in training machine learning systems. In addition to increasing the number of processors training a neural network, he realized that the speed and efficiency of those chips could also be improved. The more specifically the chips could be tailored to the kinds of calculations machine learning required, the better they would be for Ng's purposes. Most processors were akin to off-road vehicles, designed to perform a wide range of tasks. Ng wanted something closer to a sports car, specialized for AI. Though it wouldn't be able to do everything well, it would be much faster on the asphalt of machine learning.

Fortunately, just such a type of specialized computer hardware already existed. Graphics processing units, or GPUs, were then highly coveted by video gamers everywhere. In 1994, Sony announced that its PlayStation console would have a separate GPU, and in 1999, a technology company called Nvidia introduced the GeForce 256, a GPU for home computers. The GPU was designed to take the load off the CPU, which handled the more mundane tasks of running a computer. Nvidia optimized the GeForce to perform the calculations that governed game visuals, such as the shadows, colors, and lighting. It was an impressive feat of engineering that involved additional parallelization within the GPU. Computers with a GPU inside them ran games much faster and displayed them more vividly than those without.

Ng was a leader in arguing that the same GPUs that powered video games could perform machine learning calculations more efficiently than traditional CPUs.[18] Back at Stanford in 2013, he used sixty-four GPUs to train in only a few days a neural network that was six times as big as X's network, making it the largest neural network in the world at the time.[19] While the project at X had cost more than $1 million, the GPU version of it cost around $20,000.[20]

GPUs quickly came to dominate machine learning research and accelerated its progress. Researchers everywhere shifted to the more specialized hardware. Krizhevsky and Sutskever used two Nvidia GPUs to train the large neural network that powered AlexNet.[21] Likewise, rapidly improving GPU power was one of the factors that helped GANs improve so quickly, and many of DeepMind's reinforcement learning systems used GPUs to train faster. In 2014, Ng joined Baidu, the Chinese technology giant; the first thing he did there was order 1,000 GPUs for machine learning research.[22]

It was a sign of the times. The relative importance of compute versus algorithms and data remains a matter of debate, but a general pattern seemed clear to Ng: the more compute made available to train a machine learning system and the larger the neural network, the more capable the whole system would become. Many of the data- and algorithm-based successes detailed in the previous two chapters would not have been possible without the switch to GPUs.

Advances in computer chip design continued to roll out. In 2016, Google announced that it had developed what it called a Tensor Processing Unit, or TPU.[23] The TPU was a custom-built chip. Since the chip was designed to do just one thing, machine learning calculations, it was classified as an application-specific integrated circuit, or ASIC. If CPUs were off-road vehicles and GPUs were sports cars, the TPU and other ASICs for machine learning were Formula One race cars. Google said that its TPUs were fifteen to thirty times faster and thirty to eighty times more efficient than CPUs and GPUs for running neural networks, enabling more calculations per second and using less power.[24] The company had made a variety of technical improvements to TPUs that GPUs lacked. The net gains in capability were impressive.

The TPU upended the computing industry. Since its inception, Google had been a purchaser of processors, not a manufacturer; one estimate suggested that Google bought 5 percent of all the world's processors designed for computer servers, totaling more than a million devices every single year.[25] The company's move to design its own chips sent alarm bells ringing for executives at Intel and Nvidia. Some companies started building machine learning ASICs of their own, while others focused on improving chips known as field-programmable gate arrays, or FPGAs, which are powerful tools for machine learning calculations but can also be used for other purposes. Investment dollars poured in. The money showed that, in machine learning, every last bit of calculating speed and efficiency mattered.

The industry continued to race ahead. In 2017, Nvidia tried to bring the attention back to GPUs. The company announced that it was rolling out newer models that were revamped for machine learning while still being more generally useful than the focused ASICs or even the more flexible FPGAs.[26] But Google was ready for this counterpunch. The company unveiled a second version of its TPU that could perform more than a hundred trillion machine learning calculations per second, a notable improvement. Even better, multiple chips could be grouped into clusters called pods, where each TPU would work in parallel on pieces of a single large task, like training a neural network. Google said that some advanced machine learning systems could achieve high performance after just hours of training with a second-generation TPU, whereas gaining similar capability with older processors sometimes took weeks.[27] After that, the company quickly got to work developing a third version—one that performed so many calculations so quickly that it needed a special liquid-cooling system to keep from overheating.[28]

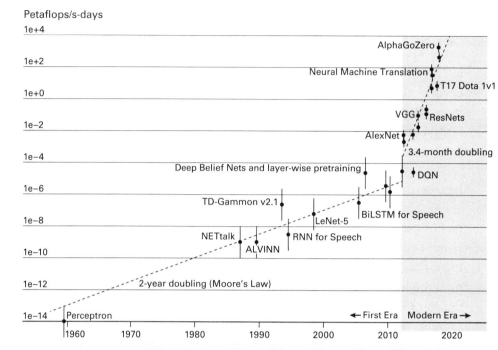

Petaflops/s-days

3.1   On this logarithmic graph, exponential growth is represented by a linear upward slope. During the first decades of computing, Moore's Law led to exponential increases in computing power, with a doubling time of around twenty-four months. After 2012, machine learning researchers were able to utilize much more computing power, resulting in an even faster rate of exponential growth, with a doubling time of just over three months.

By 2018, it was apparent that, in the six years since Ng had linked together all those processors at the X lab, a revolution in compute for machine learning had occurred. One study found that between 2012 and 2018, the processing power applied to individual advances in machine learning had increased by a factor of 300,000—doubling every three months on average.[29] The importance and power of compute would only continue to increase, and its impact would become more striking, beginning first with a virtual battlefield.

REAL-TIME AI STRATEGY

After algorithms conquered chess and Go, there was wide consensus that a video game called *StarCraft II* was the next frontier in AI versus human competition.[30] Released in 2010, the game was the leader of the real-time strategy genre and one of the most successful creations of all time. It depicts an interplanetary war with three different civilizations fighting for dominance. Playing *StarCraft II* is somewhat akin to playing a version of 3D chess in which both sides can continually make moves and need not alternate turns. The game is especially popular in Asia, where matches between leading players draw crowds of thousands to packed stadiums.

StarCraft poses a number of challenges that make AI dominance seem difficult. Most obvious is the sheer complexity of the game. Go features one type of piece per side, a simple stone; each of *StarCraft's* three competing armies deploys its own array of unique military units and structures. While Go games unfold on a board with a grid of nineteen undifferentiated squares by nineteen undifferentiated squares, *StarCraft* offers a panoply of battlefields, each with its own varied terrain. All of this complexity dramatically increases the number of possible forms each match can take. For a given decision, the player has to choose an action from among 10 to the 26th power possible options, and in total, a *Star-Craft* game has something like 10 to the 1,685th power possibilities—1 followed by 1,685 zeroes.[31] The number of seconds in the history of the universe, by contrast, is 10 to the 17th power. Even a game as complex as Go, which has 10 to the 170th power possibilities, seems like tic-tac-toe in comparison. And yet, despite all this complexity, DeepMind's Hassabis called *StarCraft* "exquisitely balanced," with no dominant strategies or unfair advantages.[32]

Not only are there more possibilities per decision, there are also simply more decisions in total. Whereas Go involves several hundred decisions per game, a *StarCraft* battle requires many tens of thousands. As in real war, players take time to develop bases, gather materials, build forces, engage in smaller skirmishes for tactical advantages, and ultimately gain a strategic edge that leads to victory. While the game is simpler and more structured than grand strategy in geopolitics, it tests similar skills of long-term planning, resource allocation, and tactical execution.

Even if AI could navigate the complex and varied possibilities, there was another significant problem to deal with: uncertainty. Go, like chess, is a perfect information game, with both players' pieces in everyone's sight at all times. *StarCraft* is not. Players can see only the area around their own army; everything else remains invisible. As a result, players must work to uncover information and make educated guesses about what forces their opponents have and what strategies they are employing. In general, *StarCraft* players have even less insight on their opponents than players in other imperfect information games, like poker (which AI mastered in 2017).[33] For an AI to win, it would have to perform reconnaissance on its opponent, make sense of the imperfect information it obtained, and then devise and implement a battle plan that balanced short- and long-term interests.

For all of these reasons, doubt persisted about AI's ability to defeat top human players at *StarCraft*. Nevertheless, in 2017, Demis Hassabis, David Silver, and the team at DeepMind began working on a project to do just that. At the time, one *StarCraft* tournament manager and computer scientist estimated that it would take five years for AI to emerge victorious. *Wired* magazine wrote at the time, "No one expects the robot to win anytime soon. But when it does, it will be a far greater achievement than DeepMind's conquest of Go"—which itself had seemed like the ultimate game challenge just two years before.[34] Even after the stunning triumph of AlphaGo and its successors, the exponential growth in what AI could do and how fast it could improve was hard to foresee; the new fire's course was clear enough, but its growing speed was still widely and easily underestimated.

DeepMind's approach to building a reinforcement learning system called AlphaStar relied on a potent mix of data, algorithms, and compute. DeepMind began by feeding AlphaStar data on how top humans played so that early versions of the algorithm could learn to mimic professional players.[35] This data was essential, since *StarCraft* was so complex that it would take too long for even the fastest AI system to discover all of the game's strategies on its own. AlphaZero had shown that, for some problems, human-provided data was not necessary; AlphaStar was a reminder that, in other contexts, data continued to matter. But data alone was insufficient. The version of AlphaStar that tried to imitate human play could defeat 84 percent of human players. It was good, but not good enough to beat professionals.[36]

AlphaStar's algorithms had to get better. At one point during its research in 2017 and 2018, DeepMind set up its own competitive *StarCraft* league, one populated entirely by different versions of the AlphaStar machine learning system. These different versions battled it out against one another, learning new strategies and uncovering which decisions were most likely to result in success. In the same way that wild cats and mice evolve over time—one getting faster at catching prey, the other getting ever quicker and nimbler—so evolved the different versions of AlphaStar. When one version learned a new strategy that was demonstrably superior, other versions adapted. The playing field would again level, and the different AlphaStars would search for the next way to gain an advantage. Using this approach, the collective versions of AlphaStar efficiently explored the vast complexity of *StarCraft*, learned how to manage uncertainty, and improved their chances of victory.

Like species evolving in a shared ecosystem, the state of play in DeepMind's league changed over time. Early on, the many versions of AlphaStar favored quick and easy strategies, such as ordering an attack on the enemy's base as soon as possible. These strategies are risky and rarely work against more advanced players. Eventually, some versions of AlphaStar learned how to defend against them, much in the way that intermediate-skill human players do. This evolution in defensive maneuvers prompted the many versions of AlphaStar to develop more mature and patient ways of playing the game.[37]

To make the versions of AlphaStar more robust, DeepMind even created variants of the system that were explicitly designed to find, probe, and exploit weaknesses in the other versions. Though the technical details differed, this approach was conceptually similar to what Goodfellow did with GANs in that it used machine learning systems to find and fix the weaknesses of other machine learning systems. It was the digital equivalent of boxers seeking out sparring partners who expose their own weaknesses and force them to improve. One of these special versions, for example, might focus on unusual styles of attack, compelling other versions to be ready for such maneuvers. Another special version might be much more patient, prompting AlphaStar to learn how to win long, drawn-out games that require more strategic planning.[38]

All of this neural network training required an enormous amount of compute. DeepMind deployed Google's third-generation TPUs.[39] Each version of AlphaStar in the all-AI *StarCraft* league used sixteen of these state-of-the-art chips while it trained. During the fourteen days the league was under way, computer chips worked around the clock, powering neural networks that played games against one another, learned from what happened, and got better.[40] Without this compute, AlphaStar would not have been possible.

The computer chips weren't cheap. Google had spent large sums of money designing and building its high-performance creations. It is unclear how much DeepMind paid to use them; as a subsidiary of Google since 2014, it likely got a discount on the chips. It would cost around $13 million for a commercial user to rent the chips for the timeframe DeepMind needed them to train the final version of AlphaStar.[41] This estimate does not include the time required to train previous versions of the code or the salaries of the people involved in getting AlphaStar to work with the chips. No matter how expensive they were, the chips paid off. In December 2018, around the same time DeepMind was entering the first version of AlphaFold into the CASP, AlphaStar beat a top human *StarCraft* player five games to zero in a secret match. Just eight months later, DeepMind unveiled three versions of AlphaStar, one for each of the three types of armies in *StarCraft*. Each version played at a grandmaster level, placing it in the top 0.2 percent of players in the world.[42] AI had surpassed another hurdle. It could win interplanetary wars—or fun simulations of them, at least.

Just as AlphaGo provided new insights into Go, and AlphaZero revealed secrets that had eluded humans for centuries in chess and shogi, so too did AlphaStar help top players explore *StarCraft*'s strategic possibilities. One expert said:

AlphaStar is an intriguing and unorthodox player—one with the reflexes and speed of the best pros but strategies and a style that are entirely its own. The way AlphaStar was trained, with agents competing against each other in a league, has resulted in gameplay that's unimaginably unusual; it really makes you question how much of StarCraft's diverse possibilities pro players have really explored.[43]

This capacity for strategic innovation had immediate relevance for warriors. They recognized that *StarCraft* was war in miniature, even more so than chess or Go, and that battle-planning machine learning systems might be able to help win real-world fights. Tactically, they noted that AlphaStar's capacity to coordinate the many units of its army could translate to the battlefield. As chapter 6 will explore, such rapid coordination among units offers new kinds of military possibilities, with robotic weapons working together to achieve objectives set by humans.

But the evangelists at DeepMind steered clear of these national security implications. Instead, the company argued that AlphaStar marked an advance for machine learning more generally. AlphaStar's ability to make short- and long-term plans would be useful in solving real-world problems that would benefit everyone. Understanding how tiny changes in actions could have far-off effects was at the core of a wide swath of challenges facing humanity, such as modeling climate change and predicting forest fires. Just as DeepMind's league had improved how AI played *Star-Craft*, other virtual competitions might help AI improve at blazing new lanes of research in science and beyond.[44]

In short, the company argued, the real-world effects of large-scale neural networks backed by massive amounts of compute were on the horizon. Just a few weeks after the first public announcement of AlphaStar, another AI company deployed its own compute-intensive system and proved DeepMind right.

ONE WORD AT A TIME

The writing prompt was simple but fun: "In a shocking finding, a scientist discovered a herd of unicorns living in a remote, previously unexplored valley in the Andes Mountains. Even more surprising to the researchers was the fact that the unicorns spoke perfect English."

The pupil seemed to grasp this story opening's many possibilities, quickly filling in some details and introducing a main character:

The scientist named the population, after their distinctive horn, Ovid's Unicorn. These four-horned, silver-white unicorns were previously unknown to science.

Now, after almost two centuries, the mystery of what sparked this odd phenomenon is finally solved.

Dr. Jorge Pérez, an evolutionary biologist from the University of La Paz, and several companions, were exploring the Andes Mountains when they found a small valley, with no other animals or humans. Pérez noticed that the valley had what appeared to be a natural fountain, surrounded by two peaks of rock and silver snow.

Pérez and the others then ventured further into the valley. "By the time we reached the top of one peak, the water looked blue, with some crystals on top," said Pérez.

Pérez and his friends were astonished to see the unicorn herd. These creatures could be seen from the air without having to move too much to see them—they were so close they could touch their horns.

While examining these bizarre creatures the scientists discovered that the creatures also spoke some fairly regular English. Pérez stated, "We can see, for example, that they have a common 'language,' something like a dialect or dialectic."

The tale continued from there, discussing the backstory of the unicorns, hinting at the possibility of extraterrestrial origins, and suggesting that a DNA test of the creatures might shed further light. It was the start of an intriguing plot, one that fit squarely within the realm of typical science fiction.[45]

But the most science-fiction element of this scenario was that the pupil writing the words was a machine learning system called GPT-2, created by OpenAI, a research lab based in California. OpenAI, like DeepMind, pursued artificial general intelligence. The lab was cofounded in late 2015 by technology titans, including Elon Musk, the entrepreneur Sam Altman, and Ilya Sutskever (of AlexNet fame). Within a few years, OpenAI had achieved some major successes. Most notably, it created a machine learning system that mastered a video game called *Dota 2*, which required strategic decision-making and coordination among a five-member team and showed how AI systems could work together.[46] But nothing would compare to GPT-2, which the lab unveiled in February 2019.

Like so many other machine learning systems, GPT-2's basic foundation was a large collection of human-generated data. OpenAI's researchers wrote a program that aggregated the text of more than eight million websites shared by the users of Reddit, a popular online community. The websites offered a large set of examples of how humans used language, providing evidence of how words combined to form sentences and sentences combined to form paragraphs. From human writing, GPT-2 learned to mimic grammar and syntax.[47]

The examples also showed how different writers in different genres used varying tones and styles. They demonstrated the way in which certain words are likely to appear together because they convey related concepts. From these examples, GPT-2 also saw how authors introduced characters to the reader and then referred back to those characters later in the text, building narratives along the way. The remit of examples was wide: Reddit's many users linked to a broad cross-section of human thought, touching on a wide variety of subjects, each becoming a tiny part of the massive data set from which GPT-2 would learn.

Next, OpenAI devised an algorithm that could process all of this data. The researchers chose to use a transformer, a type of algorithm developed in 2017 that combines both supervised and unsupervised elements. Transformers are immensely powerful, but understanding the details of how they work is not necessary to appreciate the significance of GPT-2.[48] OpenAI's transformer excelled at predicting the next word in a sentence, a fundamental test of linguistic understanding. For example, when given the prompt "The president lives in the White ___," it is easy to guess that the next word in the sentence is almost certain to be "House." White and House are two words that often pair together, especially when presidents and their residence are discussed.[49]

Guessing the next word in a sentence is not always this easy. For example, in the prompt "I moved to Berlin. Now I speak fluent ___," the missing word is most likely "German." To get this right, one must recognize that the word "fluent" is likely to precede the name of a language, that the clue for which language is being discussed appears not in the sentence with the missing word but in the previous sentence, that Berlin is a place in Germany, and that the sentences imply some cause and effect—the move to Berlin prompted the language acquisition. Most humans will immediately and subconsciously move through this chain of logical reasoning; a machine usually has a much harder time doing so.[50]

GPT-2 required a vast neural network to facilitate this word prediction. OpenAI designed a network with 1.5 billion parameters representing connections between its neurons. The machine learning algorithm could adjust each connection to give GPT-2 just a little more linguistic capability.[51] But such a large network required a massive amount of compute. To train GPT-2, OpenAI rented state-of-the-art TPUs from Google and ran thirty-two of these TPUs continuously for a week.[52]

The resulting system was remarkable. GPT-2 could credibly predict words again and again, including on text that it had itself generated, eventually adding—one linguistic brick at a time—whole sentences and paragraphs to prompts. For example, GPT-2 might predict that "I moved

to Berlin. Now I speak fluent ____" was followed by the word "German,"
and then it might keep going, producing something like, "I moved to
Berlin. Now I speak fluent German because I chat with my friends every
night at dinner."[53] With this method, the system could create whole stories, like the one about Dr. Jorge Pérez and his team's discovery of English-
speaking unicorns.

OpenAI found that GPT-2's word prediction capabilities were broadly
useful, enabling it to answer questions from a range of fields. In a test,
a researcher wrote, "Question: Who invented the theory of relativity?
Answer: ____." Drawing on its eight million pages of human data, of
which at least one must have discussed relativity, GPT-2 correctly predicted that the most likely next word was "Albert" and that the word
most likely to follow that was "Einstein." Using this method, GPT-2 correctly answered queries about the quarterback of the Green Bay Packers,
the largest supermarket chain in Britain, the Secretary of Homeland Security, the official calendar of India, and the year the first *Star Wars* film was
released.[54] The system was imperfect, especially when the questions were
more complex or the relevant answers were not in the training data, but
its success was striking.

More was to come. A little over a year later, in May 2020, OpenAI
unveiled GPT-3. The new system had a similar algorithm to GPT-2 but an
expanded data set. OpenAI began with nearly a trillion words of human
writing from vast expanses of the internet. The company's engineers then
filtered this content down to focus on higher-quality writing, and added
in additional sources of information, such as collections of books and the
contents of Wikipedia. The data set that resulted ran to several hundred
million standard pages of text.[55]

To enable this learning, OpenAI endowed GPT-3 with a massive neural
network—around 175 billion connections between neurons, more than
one hundred times larger than GPT-2—and a vast amount of compute.[56]
This approach marked something of a divergence from what DeepMind
was doing with AlphaGo and its successors, though the two companies
also had overlapping techniques. While DeepMind was increasing the
learning capability of its algorithms, enabling them to be more flexible as
a result, OpenAI was rapidly expanding the size of its algorithms' neural
networks, permitting them to process more insights from data. It was a

different pathway that OpenAI hoped would take humanity closer toward creating artificial general intelligence.

OpenAI's approach required an enormous amount of computing power. Training GPT-3's gigantic network was so mathematically intensive that it was equivalent to performing 3,640 quadrillion calculations per second every single second for a day.[57] The cost for an ordinary person to access that much compute would likely be well in excess of $10 million, but OpenAI probably got a discount from Microsoft, the company that ran the data centers powering GPT-3's training and that had invested a billion dollars in OpenAI in 2019.[58] Even so, compute was clearly a limiting factor in creating GPT-3; when OpenAI discovered a minor bug after the system had been trained, the lab decided it was too expensive and time-consuming to go back and compute everything again.[59] After studying GPT-3 and the trend lines in AI, Bryan Catanzaro, a leading industry researcher, predicted that by 2026 companies would spend upward of $1 billion on the computing power to train just one of their creations.[60]

Minor bug and massive compute costs aside, GPT-3 made GPT-2 look like a child's toy. The new version could answer science questions, correct grammatical mistakes, solve anagrams, translate between languages, and generate news stories. OpenAI tested this last ability by providing GPT-3 with a headline and a one-sentence summary. For example, when given the headline "United Methodists Agree to Historic Split" and the summary "Those who oppose gay marriage will form their own denomination," GPT-3 created the following:

After two days of intense debate, the United Methodist Church has agreed to a historic split—one that is expected to end in the creation of a new denomination, one that will be "theologically and socially conservative," according to The Washington Post. The majority of delegates attending the church's annual General Conference in May voted to strengthen a ban on the ordination of LGBTQ clergy and to write new rules that will "discipline" clergy who officiate at same-sex weddings. But those who opposed these measures have a new plan: They say they will form a separate denomination by 2020, calling their church the Christian Methodist denomination.

The Post notes that the denomination, which claims 12.5 million members, was in the early 20th century the "largest Protestant denomination in the U.S.," but that it has been shrinking in recent decades. The new split will be the second in the church's history. The first occurred in 1968, when roughly 10 percent of the denomination left to form the Evangelical United Brethren Church.

The Post notes that the proposed split "comes at a critical time for the church, which has been losing members for years," which has been "pushed toward the brink of a schism over the role of LGBTQ people in the church." Gay marriage is not the only issue that has divided the church. In 2016, the denomination was split over ordination of transgender clergy, with the North Pacific regional conference voting to ban them from serving as clergy, and the South Pacific regional conference voting to allow them.[61]

OpenAI showed this story and other GPT-3 stories to a group of human readers and asked them if the stories had been written by a human or by a machine. Only 12 percent of readers correctly said that a machine had written the story about the Methodist split. For religious history experts, there was an easy tell: the Evangelical United Brethren Church had in fact merged with the Methodist Church in 1968, rather than split from it. GPT-3's text also was unclear about whether the new denomination would support gay marriage or oppose it (as the prompt had stipulated). Notwithstanding these errors, GPT-3's writing was remarkably convincing to most readers. Even with GPT-3's worst-performing news story—a jumbled tale about Joaquin Phoenix's tuxedo at an awards show—only 61 percent of readers guessed that the text was machine-authored.[62]

GPT-3 could undeniably keep up with humans in a wide range of arenas. Variants of GPT-3 passed muster with expert communities, such as one version that learned how to prove mathematical theorems well enough that a major formal mathematics library accepted its proofs.[63] Another version coded websites in response to human commands.[64] Still another version used GPT-3 to generate not text but images that met a specific description; a user could type instructions to generate "a female face with blonde hair and green eyes" or "a picture of an armchair shaped like an avocado" and the machine learning system would do so on the fly.[65]

Machine learning researchers debated the significance of the advances in language processing that powered GPT-2 and GPT-3. One expert dismissed the word prediction technology, writing after GPT-2's unveiling that it "is a brute-force statistical pattern matcher which blends up the internet and gives you back a slightly unappetizing slurry of it when asked."[66] Another expert argued that GPT-3 was too dependent on its vast neural network and the compute required to run it and that it lacked a fundamental scientific advance, while still others mocked it as a "blovia-tor."[67] In this view, machine learning's capacity to produce pseudorealistic imitations of human writing, even laced with real-world facts or semi-narrative structures, was unimpressive. The systems' results more closely resembled plagiarism than insight, and no amount of compute could change that.

Others took the opposite, and more optimistic, view. They argued that language prediction was a meaningful step toward artificial general intelligence, though much more work remained to be done. Rather than plagiarizing ideas, GPT-3 showed that AI could learn from human data to derive new insights, just as AlphaGo had devised move 37 against Lee. In this view, word prediction was a pathway to solving any problem that could be rendered as text, from science to poetry to translation; indeed, GPT-3 showed promise at all of these pursuits, despite the fact that it had not been designed to perform them. Scott Alexander, a well-known pseudonymous science commentator, thought that much more was possible: "Learning about language involves learning about reality, and prediction is the golden key," he wrote. "'Become good at predicting language' turns out to be a blank check, a license to learn every pattern it can."[68] Whereas AlexNet, AlphaGo, and AlphaFold had shown the dominance of AI in discrete tasks, the GPT breakthroughs showed the increasingly general capacity of AI systems—making it easier still to imagine their applications to national security.

The improvement in machine language prediction abilities, fueled in large part by exponential growth in the amount of computing power applied to AI, shows no signs of slowing. This continued progress in computer hardware is a technical marvel; for decades, semiconductor pioneers have kept the pace of innovation high, making ever-better machines. But the implications of their actions stretch far beyond the world of computer chips. Through the story of one of the semiconductor industry's leaders, we can see how shifting technical frontiers acquired enormous geopolitical importance for democracies and autocracies alike.

A GLOBAL INDUSTRY

Zhang Zhongmou—known in the West as Morris Chang—was born just outside of Shanghai in 1931. Like Ian Goodfellow, as a child Chang dreamed of becoming a novelist.[69] He spent large portions of his childhood moving across China as his family fled the violence of the Sino-Japanese War, World War II, and the war between Chinese nationalists and Communists.[70] After he finished high school, he moved to the United States and became the only Chinese freshman at Harvard University. America quickly made an impression on him. "What a country!" he said later. "The United States was at its peak in its moral leadership, in its political leadership in terms of democracy—and it was the richest country in the world." At university, Chang's American Dream began. It was an experience he later likened to "sheer ecstasy."[71]

Reality soon set in. Being Chinese, Chang felt that only a few career paths were open to him. "There were Chinese laundrymen, Chinese restaurateurs, Chinese engineers, and Chinese professors," he said. "Those were the only respectable professions for Chinese—no lawyers, no accountants, no politicians."[72] He gave up on his dream of becoming a writer and decided to become an engineer. Chang transferred after a year to MIT in pursuit of a more technically rigorous education, earning his undergraduate and master's degrees in mechanical engineering in 1952 and 1953, respectively. His academic career seemed to end abruptly afterward, when, in pursuit of a PhD, he failed his qualifying exams; he said later that he just didn't study enough.[73]

With no PhD in sight, Chang entered the job market. In 1955, the automotive industry was a bedrock of American business, and Ford offered him a job for $479 per month. But a company called Sylvania Electric Products offered Chang $480 to work in a small division making transistors and semiconductors, which had been pioneered just a few years before. Chang asked Ford to beat the slightly higher offer, but they turned him down. It was a fateful decision that would echo for decades, though no one knew it yet. Chang accepted Sylvania's offer and got to work in the new industry.[74]

At the time when Chang joined Sylvania, the United States was leading the semiconductor industry. Chang excelled at the business of designing these transistors and the computer chips that used them. After three years at Sylvania, Chang moved to Texas Instruments, a leading electronics company.[75] With his writing aspirations shelved, Chang's new dream was to become vice president of research and development for the firm, but his bosses told him he would need a PhD.[76] At age thirty, he went

back to school, enrolling in the doctoral program in electrical engineering at Stanford. Texas Instruments agreed to foot the bill and pay his sal-
ary while he studied.[77]

It was a good investment for both the company and for Chang. After
getting his PhD—this time passing his qualifying exam with "flying
colors"—Chang came back to work.[78] He worked at Texas Instruments for
twenty-five years, eventually leading its worldwide computer chip division. But the company began focusing more on consumer products like
calculators and toys. Business was mediocre, and Chang was put in a
vague role overseeing "quality and people effectiveness."[79] He thought he
could do better. In 1983, he left.[80]

Chang first tried a job in New York, as president and chief operating
officer of General Instruments. Things quickly turned bleak. The company wasn't as serious about research and development as he wanted,
and he disliked the New York weather. He and his wife separated.[81] His
prospects of leading an American technology company seemed dim. Venture capital was all the rage at the time, but Chang didn't think he'd be
very good at it, so he looked for other options.[82]

Around this time, a Taiwanese government official named K. T. Li
approached him. Chang had met Li a few years before when he was at
Texas Instruments, but this time Li had something intriguing to offer:
Taiwan wanted to revitalize its Industrial Technology Research Institute, a
decade-old facility that was the most significant research lab in the country.[83] Would Chang leave New York and lead it? Everyone told Chang not
to do it, pointing out that it would end his dream of being an American
executive. He said yes nonetheless.

Chang quickly made enemies at the lab. To Taiwanese engineers, he
was a brash American with a different and more aggressive way of doing
things.[84] He instituted stricter performance reviews and put employees on
probation. He earned a reputation as someone who didn't "tolerate fools"
and threatened to fire those who didn't meet his standards—a common
threat in American business but deeply unusual in Taiwan at a time when
government employees were rarely dismissed.[85] Taiwanese legislators
received hate mail criticizing him.[86]

Just a few weeks after Chang joined the lab, Li approached him again.
Taiwan's government wanted to grow its semiconductor industry.[87]

American companies like Texas Instruments, Intel, and Advanced Micro Devices were leading the market, but Taiwan thought it could compete. Even though the country was far behind in technical capability, Chang felt he had to accept Li's proposal or else he would be seen as lacking ambition. His future in Taiwan depended on succeeding at this task, and to turn down the opportunity was to surrender without trying. "It's like in the movie *The Godfather*," he said, "an offer you can't refuse."[88]

In 1987, Chang left the lab and founded Taiwan Semiconductor Manufacturing Corporation, or TSMC. He knew that his team didn't have the expertise to design computer chips from scratch. Instead, he developed what came to be known as the foundry model, in which TSMC took other companies' chip designs and brought them to life in its giant fabrication facilities.[89] Over the course of almost two decades, Chang—affectionately known as the "Foundry Father"—led TSMC in opening a dozen of these foundries, each producing more advanced chips than the last through increasingly intricate techniques.[90] Time and again, Chang and TSMC made major bets on cutting-edge technology to improve the fabrication process and successfully gained market share as a result. The world took notice and, in 1996, TSMC became the first Taiwanese company to be listed on the New York Stock Exchange.[91]

TSMC's business evolved further after Ng, Dean, and others showed how massive computing power could help AI systems work more powerfully than ever before and how important computer chips were to making it happen. Chang led the company to embrace the importance of machine learning. TSMC churned out chips designed by some of the leading firms in the world, including Apple and Nvidia.[92] It built fabrication facilities that cost upward of $20 billion, exponentially improving the processing power of its products.[93] With business booming, Chang at last stepped down in July 2018, a month shy of his eighty-seventh birthday.[94]

The rise of TSMC—and the semiconductor industry more generally—is a story not just about business or technology but about geopolitics; Chang himself has explicitly acknowledged as much.[95] Even before TSMC achieved success in Taiwan, some Chinese leaders wanted to compete in the semiconductor industry and eventually began investing large sums in an attempt to develop a competitive domestic manufacturing base. These nascent efforts suffered from unrealistic expectations; in 1977, some planners imagined a semiconductor manufacturing base could be created in a year.[96] In fact, several decades would have been a more accurate estimate, since it was nearly impossible to overcome the head start enjoyed by other countries when combined with the exponential growth in chip performance as governed by Moore's Law. China's continual difficulty with recruiting top semiconductor manufacturing talent also presented a major barrier.[97] A series of aggressive industrial plans from Beijing failed to close the gap, as China learned the challenges of the intricate business the hard way.

In the early 2000s, things got better for China. Some of the people who had gone overseas in the 1980s and 1990s returned to China with electrical engineering skills developed at American universities and companies. They put this hard-won knowledge to work, starting an array of new businesses in China focused on different parts of the semiconductor manufacturing process, including companies like Semiconductor Manufacturing International Corporation and HiSilicon that would become behemoths. Even as government subsidies waned, these firms were able to keep China's domestic industry moving forward, though they still lagged substantially behind competition like TSMC.[98]

In 2014, the Chinese government again began making major investments through its China Integrated Circuit Industry Investment Fund, more commonly known as "the big fund."[99] These investments supported research and development efforts by China's leading chip manufacturers, boosted top engineers' salaries, and aspired to reduce China's dependence on imports. In addition, as part of its Made in China 2025 effort that formally launched in 2015, the Chinese government began to invest what it said was more than $100 billion in semiconductor manufacturing. The effort aimed to meet 80 percent of domestic demand for chips by 2030.[100] To complement these moves, in 2017, China announced the New Generation Artificial Intelligence Development Plan, which, among other objectives, included ambitions to improve China's position in the computer chip industry.[101]

Thus far, the results have been mixed, despite China's lavish state subsidies and aggressive intellectual property theft.[102] China has as yet produced no chip design software that is globally competitive. It has made some progress with designing its own chips using imported software, but

American companies dominate the market. What's more, China's fabrication facilities are not competitive with TSMC's or those of other companies; Chang estimated in 2021 that the mainland's facilities were five years behind TSMC in terms of manufacturing technology.[103] In 2020, 84 percent of chips sold in China were made elsewhere—about the same percentage as in 2014—and even those produced in-country were made by non-Chinese companies more than half the time.[104] Top Chinese chip companies, such as Horizon Robotics and Cambricon, still outsource chip manufacturing to TSMC. As for the chip manufacturing equipment within fabrication facilities, five companies in the Netherlands, Japan, and the United States control the vast majority of the world market that produces such equipment, and China is not in the mix.[105] As a result, China must import more than $300 billion in computer chips each year.[106]

Computer chips thus offer one of the fundamental democratic advantages in the age of AI. This advantage stems in large part from where key talent, like Morris Chang, emigrated decades ago, as well as from savvy government investments that provide benefits to this day. Democracies benefit as well from computing power's role as the third spark of the new fire—a vital, though often-overlooked, part of advancing what AI can do. But before we turn in part II to all the ways in which compute-intensive AI systems could change the world, it is worth considering the ways these systems might fall well short of expectations. Even with the best computer chips, the dangers of failure still loom large.

4

FAILURE

In 1954, a group of well-known AI scientists and linguists from Georgetown University and IBM made a bold prediction: in three to five years, computers would be able to translate between languages as fluently as humans could.

The statement came just after the group's demonstration of machine translation at IBM's New York headquarters. The researchers had taken an IBM 701, a giant computer, and programmed it with six very basic rules of grammar and 250 word stems and endings.[1] To test its translation ability, they gave it sixty Russian sentences, carefully choosing sentences that wouldn't pose too many challenges for the machine. They selected organic chemistry as a specialty to show off the machine's scientific ability, as well as more general subjects to demonstrate its flexibility.

The results seemed remarkable. "Iron is obtained from ore by chemical process," read one sentence the machine translated from Russian. "International understanding constitutes an important factor in decision of political questions," read another.[2] More work lay ahead, but this demonstration marked, according to the IBM press release, the "Kitty Hawk of electronic translation," a reference to the North Carolina site of the Wright brothers' first flight.[3]

Newspapers all over the country put the story of the Georgetown-IBM experiment on the front page. The *New York Times* noted that the machine had a small vocabulary, but assured readers that great progress was just ahead. "There are no foreseeable limits to the number of words that the device can store or the number of languages it can be directed to translate," the paper wrote. The *Christian Science Monitor* was similarly effusive, declaring, "The 'brain' didn't even strain its superlative versatility and flicked out its interpretation with a nonchalant attitude of assumed intellectual achievement."[4]

Even though the results were good, the process was unimpressive in hindsight. To derive these translations, the system did little to parse the construction of the sentences it received, such as identifying the grammatical structure in much depth. Instead, the methodology was more akin to using a dictionary to look up individual words. Above all, the experiment worked because the given sentences all used words in the machine's vocabulary.[5] By today's standards, this kind of machine translation is laughably limited. By the standards of the 1950s, this experiment and other early AI efforts were enough to ignite a wave of extraordinary hope, hype, and confidence that major success was just a few years away; Claude Shannon, one of the most important computer scientists ever, predicted in 1960 that "within a matter of 10 or 15 years, something will emerge from the laboratory which is not too far from the robot of science fiction fame."[6]

AI researchers linked their technical ambitions to national security concerns from the start. The IBM-Georgetown demonstration's organizers deliberately chose to translate from Russian to English as a way of attracting the attention of the federal government. US national security officials took the hint and immediately recognized what the prospect of rapid machine translation could do for them. A computer that could read Russian could better track Soviet scientific progress, more quickly analyze intelligence intercepts, and facilitate communication between the two sides' leaders.

Money to build such a translation device poured into IBM and other firms from the Department of Defense, the CIA (routed via the National Science Foundation to preserve secrecy), and other parts of the government. The Soviet Union learned of the experiment and began a parallel effort to translate from English to Russian, with funding from the KGB intelligence service. In 1956, the Soviet creation possessed a vocabulary of around a thousand words in each language.[7]

Despite both superpowers' investments, the prediction of near-term success was wildly incorrect. Some machine translation systems made their way into operational use, but their accuracy remained limited. The exponential growth in capability that the evangelists of machine translation had foreseen did not materialize, as it proved fiendishly difficult to replicate in machines humanity's innate capacity to parse and deploy language. A simple example illustrates the challenge of disentangling the meaning of words in context: when translating the biblical verse "The spirit is willing, but the flesh is weak," a machine could plausibly spit back something akin to "The whiskey is strong, but the meat is rotten."[8] No human translator would make a similar mistake.

By 1964, ten years after the Georgetown-IBM experiment, the US government grew concerned about the lack of progress. It organized a committee to study the matter. For two years, the committee evaluated the original system and its successors. Its conclusion was stark: it saw "no immediate or predictable prospect of useful machine translation."[9] As a direct consequence of the report, research funding for machine translation in the United States cratered for almost a decade; it would not be until four decades later that the field would deliver on at least some of its early promises.[10]

Even this short glance at past failure amid AI hype should yield humility about the current wave of machine learning. The historical pattern is clear: many excited pioneers routinely overestimated their ability to spot and surmount significant technical roadblocks and underestimated the time it would take to do so. Often buoyed by comparatively narrow triumphs, like the Georgetown-IBM experiment, they forecasted exponential progress that never came.

This history of failure feels distant at a moment when machine learning technology is ascendant in so many areas. Today is different, the argument goes.[11] But the data does not fully support such optimism. Many, though certainly not all, of the successes of machine learning thus far have been in comparatively narrow areas. The technology often does not account well for variation in real-world environments, where machine learning systems fail more easily than in training. Even for areas of comparative success, like image recognition, machine learning systems lag behind the standards of reliability established for mission-critical systems, like aviation software. The Cassandras believe that these well-documented disparities should prompt caution before using machine learning for vital tasks, especially because it is hard to predict real-world failures from tests.[12]

Perhaps most striking of all, machine learning systems today fail differently than previous attempts at AI did. Today's systems can exacerbate preexisting biases while coating their decision-making in a façade of impartiality. The systems are opaque, sometimes producing correct answers but almost never offering supporting logic. Worse still, these systems sometimes interpret their creators' instructions far too literally and possess little to no understanding of the task at hand. All of these failures could carry geopolitical consequences and could compound into more significant harms; when the new fire gets out of control, its exponential force poses serious risks.

Much of the debate in the chapters to come revolves around the interplay between machine learning's seductive success and its embarrassing shortcomings. It is easy for policymakers to look at the machine learning breakthroughs described in the last three chapters and imagine what the technology can do for them; it is harder, but necessary, to give sufficient forethought to what the technology can't do and to consider ways in which narrow success might breed undeserved hype.[13] The failures of the past offer needed context for the excitement of the present and should temper our vision of the future. And so, with history and geopolitical implications in mind, let us consider head-on the ways in which today's machine learning can fall short.

BIAS

Amazon was an early pioneer in machine learning. After Ng's experiment and AlexNet's triumph in 2012, Amazon put the technology to greater use all across the company. It deployed advanced robotics in its warehouses, improved prediction in its logistics chains, and bought promising AI startups. Beginning in 2014, the company also applied machine learning to its hiring practices, which one employee called "the holy grail."[14]

Amazon wanted to build a machine learning system that would find the best employees among its job candidates. The company was in the midst of a gigantic expansion, during which it would add hundreds of thousands of people to its payroll.[15] It needed a way to sift through the many résumés that came in every day. Humans would continue to be involved in the process of interviewing and selecting candidates, but if machine learning could help spot talent, it would give Amazon a huge competitive advantage over other companies less able to identify good hires.

Amazon decided to use supervised learning to solve this problem. As training data, Amazon's engineers fed the system ten years' worth of résumés and hiring decisions. In the same way that the ImageNet data set provided insights for AlexNet and other systems, this data set showed the company's machine learning system what kinds of candidates Amazon had hired in the past so that the system could identify similarly promising candidates in the future.

After a year of training and development, Amazon's machine learning system had, among other things, found one particular pattern across the company's hires: they were disproportionately men. The system discerned that, in the ten years of hiring data it had scrutinized, Amazon's managers had routinely given preference—perhaps without knowing it—to male candidates over female ones. The system learned to mimic this practice in its own classification of which candidates were worth hiring, just as it had learned to extract other patterns from the training data it had been fed. As a result, an Amazon audit found that the system learned to penalize candidates who included the word "women" on their résumé (for example, if they captained a women's sports team or club) and that it systematically downgraded the graduates of two all-women colleges.

The machine's decisions serve as a reminder that supervised learning systems learn all that they will ever know from training data. These systems have no way to know which patterns are worth mimicking and which are undesirable. It is challenging to instruct a machine learning system to learn only particularly desirable patterns from data and to avoid learning things that have negative social consequences, like racial or gender bias. If a data set reflects biases in the real world, machines will usually learn these biases.

Attempting to find a way around this problem, Amazon's engineers hardcoded rules into the system that would instruct it to ignore any explicitly gendered language. Even this was not sufficient to guard against gender bias, however, since they could not determine if the machine learning system would use other less obvious means to give preference to men in its recommendations. For example, the system, like human decision makers at Amazon before it, came to prefer candidates—typically men—who used more dynamic language on their résumés, such as the words "captured" and "executed."[16] Amazon's system was not an outlier in using factors that strongly correlated with gender; another company's machine learning recruiting system learned to prefer candidates who played high-school lacrosse—a game largely played by men.[17]

The problem was insidious. Given the persistent risks of gender bias and the general failure of the system to consistently provide good recommendations, Amazon's executives eventually soured on the idea of using machine learning in hiring. Around 2017, they pulled the plug on the project.[18] It is unclear how much harm the system did while it was in development and use. Amazon employees told *Reuters*, the news outlet that broke the story, that recruiters at the company had looked at the tool's recommendations as one of several factors in their decisions. The company said that the system was not used to evaluate candidates, but did not deny that recruiters accessed its rankings.[19]

The Amazon case illustrates how challenging it is to overcome bias in machine learning, especially when the past actions of potentially biased humans serve as training data. The pattern seen in the Amazon case is visible elsewhere: a machine trains on a record of biased decisions, learns biases from that data just as it learns other patterns, and then makes biased judgments itself. Sometimes, the bias in data sets can go overlooked for years. In 2020, MIT and New York University took offline a data set of almost eighty million labeled images, drawn from the internet, that taught machines racist and misogynistic slurs; the data set had been available since 2008 and was cited in more than 1,800 machine learning research papers.[20] Even when machine learning engineers are more alert to the issue, it is still extremely difficult to stamp out subtle and implicit bias, as Amazon eventually realized.[21]

Bias in machine learning can also originate with training data sets that are incomplete or nonrepresentative. Machine learning systems generally struggle to extrapolate patterns from small amounts of data. If certain groups of people, such as people of color, are not well represented in the training data, a machine learning system that trains on that data is likely to perform substantially less effectively with those groups, and may even discriminate against them. For example, a major training data set used— among other places—in a landmark facial recognition breakthrough by Facebook in 2014 disproportionately contains pictures of White men, and it is likely that systems trained on that data set will exhibit biases as a result.[22]

Another potential cause of bias is that the developers of machine learning systems are themselves disproportionately White and male. At Facebook, 15 percent of AI researchers are women, while at Google women make up 10 percent. At top AI conferences, women author only 18 percent of the accepted papers, and 80 percent of AI professors are men. Across the board, Black people are a tiny minority in AI, usually with a percentage of representation in the low single digits, and they have often been kept out of the development of companies' AI strategies and have found their concerns ignored. In 2020, Timnit Gebru, the woman who co-led Google's ethical AI team and was one of the company's leading Black researchers, coauthored a paper that pointed out some of the drawbacks, including bias, of a great deal of research around AI and language. In response to an internal dispute about the paper, and perhaps other disagreements about the company's direction, Google fired Gebru.[23] The case offers additional evidence for the view, shared by some leading experts, that AI engineers and executives worry less about bugs or performance weaknesses in machine learning systems that primarily affect people who do not look like them.[24] As a result, these engineers and executives are less likely to fix those failures, and so biases in machine learning systems are more likely to persist.[25]

It can be hard to disentangle these various sources of machine bias from one another, but easier to see the cumulative effect. From how data is collected to the goals used to optimize algorithmic performance and beyond, one simple fact remains: machine learning systems routinely discriminate. Facial recognition systems provide an obvious and troubling example. Many researchers have expressed concern over landmark studies showing that major machine learning–based facial recognition systems perform substantially worse on non-White faces.[26]

The most well-known of these studies is a test by the National Institute of Standards and Technology, a science agency in the US government. Building on the work of academic researchers, it examined 189 facial recognition systems and found that they had a false positive rate that was often between ten and one hundred times higher when asked to identify the face of a Black or Asian person than when asked to identify the face of a White person.[27] This is a staggeringly high rate of failure, one confirmed by other studies, including a test by the American Civil Liberties Union that showed that Amazon's facial recognition system matched twenty-eight members of Congress to criminal mug shots, with a disproportionate number of the false matches pertaining to members who are people of color.[28] The harms wrought by failures in facial recognition can lead to grave civil rights violations; in the United States, innocent Black people have been arrested and jailed solely on the basis of such flawed systems.[29]

These facial recognition failures show the dangers of machine learning bias in the context of domestic affairs, but the risks extend to the context of international security too. Machine learning systems trained to spot anomalous behavior, analyze intelligence, or fire weapons will all at times be subject—absent significant intervention and technological improvements—to the same kinds of bias that caused the aforementioned systems to fail. Given that national security often involves interacting with citizens from other countries, failures due to bias deserve significant attention, if not for ethical reasons, then certainly for practical ones. The risk is hardly hypothetical; in 2019, police using a facial recognition system mistakenly identified a suspect in a terrorist attack.[30]

Unfortunately, the problem of machine learning bias is arguably even harder to solve in international security than in a domestic context. The training data for machine learning systems might be sensitive intelligence. It is highly unlikely that such information is as well-structured and straightforward as Amazon's collection of résumés and hiring decisions, making it more likely to be imperfect or incomplete. In addition, classification concerns will likely restrict who sees that data, perhaps making it harder for anyone to interrogate what the machine is learning and how it is performing.

This secrecy will likely amplify one of the final problems when it comes to machine learning bias: unless one is actively looking for it, bias is hard to find. The Amazon case is a relative success, at least insofar as the company found the system's faults and discontinued its use. The only thing worse than finding the bias in a machine learning system is *not* finding it and harming people as a result. In 2019, researchers found that an algorithm that had been used for years to manage the healthcare of millions of people in the United States had systematically underestimated the needs of the sickest Black patients, even though the algorithm itself was explicitly programmed to ignore race. The algorithm reduced the number of Black patients who received extra care by half, led to less healthcare spending overall on Black patients with the same level of need as White patients, and falsely determined that sick Black patients were healthier than equally sick White patients.[31]

The evangelists and Cassandras sometimes diverge when faced with failures like these. For evangelists, just because an AI system yields some bad outcomes does not mean that it must be abandoned.[32] In addition, because bias often comes from the training data sets as opposed to algorithms, the issue in their view is not with machine learning itself but with humans. Pedro Domingos, a widely cited AI researcher, wrote that "there is not a single proven case to date of discrimination by machine learning algorithms."[33] For Cassandras, the distinction between data and algorithms is less important than how the overall system performs; if machine learning systems act in a biased manner, their bias is a reason to hesitate in deploying them, no matter the cause.

Worse, Cassandras argue, is that computers often provide a veneer of impartiality that is not to be trusted. This veneer cloaks machine learning systems' biases in precise—though sometimes inaccurate—mathematical processes rather than hatred or explicit discrimination. It can lead to what is sometimes called bias laundering, in which machines appear to be fair but in fact are not.[34] These persistent failures cause some Cassandras to question whether machine learning is simply inappropriate for some decisions in which it could disproportionately hurt less empowered people.[35] The evangelists reply that, given how biased humans can be, algorithms could offer an opportunity to improve fairness if they learned to act more impartially than humans.[36] With the right incentives in place, they argue, such a goal is achievable.

Perhaps machine learning systems would be more trustworthy and their biases more apparent if they could explain how they reached their conclusions. Sadly, that is another area of significant failure.

OPACITY

Admiral Grace Hopper was one of the most esteemed computer scientists of the twentieth century. As World War II broke out, she tried to leave her job as a mathematics professor at Vassar College and join the Navy, but she was rejected because of her age and small stature. Instead, she joined the Naval Reserve and was posted to Harvard University. There she served as one of the operators of the Mark I, an early electromechanical computer. After the war, Hopper continued to develop computers and helped invent languages, like COBOL, that were both readable by humans and easily convertible to a form that machines could process. She helped make computing more transparent to its users and creators, and it is no exaggeration to say that some of the core principles of modern software development rest on her insights.

Less storied than her other accomplishments, though, is an incident that occurred in 1947, when Hopper was working on the larger Mark II computer. One day Hopper and her team noticed that the machine seemed erratic. As part of their investigation, they opened the casing and found a moth stuck inside that was interfering with one of the electrical relays. They carefully extracted the moth from the machine and taped the dead insect to a logbook in their written report on the error. Afterward, the Mark II worked again, and the term "bug" became more prominent in the computer engineering lexicon.[37]

Bugs are inevitable, though in today's computer systems they are figurative rather than literal. It is a simple fact of software development that things will go wrong, assumptions will no longer hold, data formats will change, typos will happen, hardware will fail, and programs will crash. No one in the decades-long history of computer science has designed and written perfect code all of the time. The art of debugging and quality assurance lies in finding these failures and remediating them before they cause major problems.

The essence of the traditional debugging process is to make the code's design and behavior transparent. Testers review the code programmers have written, often line by line, to understand what it is supposed to do and to confirm that it fulfills those intentions. They run the code in different conditions to see how it performs in practice, connecting its success or failure back to the way it is designed and built. They ensure that the code's logical sequencing is documented well, so that its actions are clear and predictable. The result is a system that is as bug-free as possible and in which users can have confidence.

Machine learning upends all of this. Machine learning systems are far less transparent than conventional computer programming or expert systems. Humans continue to make major design decisions, such as which data sets to use, which algorithms to deploy, and how much computing power to provide, but much of the system's behavior is learned by the machine itself in a way that is remarkably opaque. The distributed calculating power of a neural network is wonderful for teasing out patterns and finding the right answers to complex problems, but it is extremely poor at explaining how it derives those answers. With sometimes several hundred billion connections between neurons, it is difficult to interpret what each neuron is doing and which parts of the training data are most important to the network's performance.[38]

This challenge calls to mind advice from the famous science fiction editor John Campbell. When Campbell's writers needed inspiration, he would instruct them to come up with a creature that thinks as well as a human, or even better than a human, but not exactly like a human. He argued that it was this rare capacity, to reason differently but no less effectively, that would enliven imaginations and excite readers; *Star Trek*'s Spock and *Star Wars'* C-3PO seem to be characters who demonstrate Campbell's point.[39] Neural networks, with their different ways of processing information, opaque functioning, and phenomenal facility with certain tasks, are perhaps the closest real-world fulfillment of Campbell's challenge. An interdisciplinary group of scientists published a notable article in *Nature* in 2019, arguing that, instead of studying machine learning systems using traditional logical or analytical methods, researchers should study their actions as they might examine animal behaviors—a reminder of just how different and inscrutable machine learning systems are from traditional computer programs.[40]

As a result of this opacity, machine learning systems are often designed iteratively. Designers will try many configurations of the neural network with one set of training data and see which works best in simulations or tests. They will then tweak that version again, adjusting and readjusting the big-picture design of the system, such as the number of neurons or how fast the network learns. These design choices shape how the neural network adjusts the connections between its neurons when it trains, which in turn determines how well it performs. Even experienced machine learning engineers sometimes struggle to predict in advance which configurations will work best and to explain after the fact why some designs perform well and some others do not.[41]

The net effect of this lack of transparency is twofold. First, when machine learning systems fail, it can be very challenging to figure out why and what to do about it—unlike with the bug Hopper taped to a logbook or the virtual bugs that are visible in code written in traditional computer languages. Sometimes machine learning systems fail because of the random initialization of their neural network; run the algorithm once and it works, run it a second time with the same code and it fails for no discernible reason.[42] At times, they fail because of human error, such as typos or poor designs, or because of structural issues, like biased training data. Or these systems fail because they encounter novel environments or situations for which their training data offered insufficient preparation. Disentangling which failures come from which causes is much harder when there is so little transparency and when the systems are so complex.

Second, even when machine learning systems do succeed, the lack of transparency can and perhaps should reduce users' trust in AI. As on a high school math test, sometimes just getting the solution is not enough; showing the work matters too. For high-stakes applications that deeply affect people's lives—like virtually all possible applications to national security—some Cassandras believe that it is just as important that a machine learning system be able to explain its reasoning as it is to ensure the outputs of such a system are accurate. This transparency would make it easier to trust the system and would permit its performance to be more thoroughly audited and understood.

A case from the 1973 Arab-Israeli conflict shows the importance of trust in military applications. The Israeli Army was forced back on its heels after surprise attacks by both the Egyptian and Syrian militaries. The United States had supplied Israel with advanced TOW missiles, which adjusted their flight paths to hit targets identified by human operators. The US Army had trained with these missiles and had come to trust them; the Israeli Army had no such training and no such trust. As a result, even though the missiles would have greatly helped to repel further Egyptian and Syrian attacks, Israeli soldiers were reluctant to fire them.[43] The same reluctance might inhibit some of the uses of future AI systems unless operators can train with them and understand how they work under different conditions.

The lack of transparency is a significant barrier to using and trusting machine learning. In the absence of an explanation, it is a challenge to avoid the twin perils of not trusting enough and trusting too much. In one 2016 study, researchers put forty-two volunteers in a simulated fire emergency.[44] A robot was on hand to guide them safely to the exits, though it provided no explanation to the humans for its decisions. The only problem was that, as part of the experiment, the robot made a series of errors that would have proved fatal in a real-world scenario—and yet nearly 90 percent of the participants continued to follow the machine as it passed the exits and led them down a series of wrong turns and blind alleys.

To build public trust in AI and avoid insidious harms, some Cassandras have argued that there is a "right to an explanation." They believe that certain decisions—such as approving or denying loans, and sentencing for criminal defendants—are so important that they should not be given over to an opaque machine learning system, even if that system seems to perform better than more transparent alternatives. What sorts of explanations are sufficient is a subject of intense debate.[45] In the chapters to come, Cassandras will make similar arguments about the use of machine learning in the context of national security, contending that the stakes are too high for decisions to be left to machines that cannot explain why they do what they do.

The debate isn't going away. The technological trend lines indicate that these challenges might get harder, not easier, even as machine learning improves and teams of humans and machines become the norm, not the exception. Many AI engineers use a newer and more powerful technique called neural architecture search, in which machine learning systems design other machine learning systems; in effect, these systems are black boxes that build black boxes inside themselves. While this approach can often improve the overall capability of the resulting system and shorten the time it takes to solve harder problems, it further reduces engineers' ability to understand how the system works.[46] Absent an unexpected breakthrough, machine learning is poised to become still more opaque.

SPECIFICATION GAMING

In 2016, the staff at the company OpenAI—the creators of GPT-2 and GPT-3—were working on developing a reinforcement learning algorithm that could play video games. This work occurred several years before DeepMind's success at *StarCraft*, and OpenAI's team chose a much simpler game called *Coast Runners*. In this game, players race one another in boats as they drive around a course, earning points for driving through green targets that appear throughout the waterway. It is a concept simple enough for most schoolchildren to grasp: get to the finish line while hitting as many targets as possible along the way.

Every reinforcement learning system, or agent, seeks to maximize a specified outcome. OpenAI specified that the *Coast Runners* agent should try to maximize the score in the game, a straightforward objective. The lab's scientists set the agent loose and let it play for a while. They knew the agent would try out a variety of tactics to see which ones yielded higher and lower scores, and that over time it would gravitate toward choosing those tactics that brought better results, learning how to win at *Coast Runners* in an iterative fashion.

After OpenAI's reinforcement learning agent had trained enough to produce high scores, its creators decided to see its tactics for themselves. What they found surprised them: the agent wasn't even racing toward the finish line. Instead, while all of its competitors zoomed on ahead, the

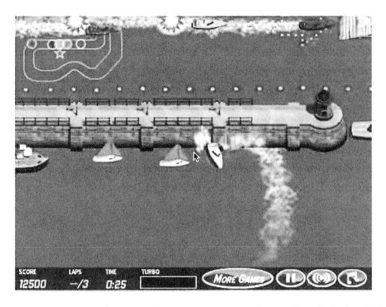

SCORE | LAPS | TIME | TURBO
12500 | —/3 | 0:25

4.1 OpenAI's reinforcement learning agent trained enough to find a loophole in the game *Coast Runners*. Even though the game was supposed to be a race, the agent pulled its boat over to a small lagoon where it could drive in circles—sometimes crashing and catching fire—and hit three targets that continually reappeared, thereby maximizing its points.

agent pulled its boat over to a small lagoon on the course, where it commenced driving in circles. It looked like an embarrassing failure.

It wasn't, though, at least not entirely. The agent had discovered through its trial-and-error training process that in this lagoon, three targets would appear. By maneuvering its boat into a weird loop, it could hit them all easily. By the time the agent came around for another loop, the three targets would appear again, and it could hit them again. And so the agent went, driving its boat in circles, going the wrong way, crashing into barriers and walls, causing its boat to catch on fire—but also hitting three targets every few seconds, and thus maximizing the score it was told to maximize.

The agent had both succeeded and failed. It succeeded because this strategy resulted in scores that were about 20 percent better than the ones human players obtained, even though the humans finished the race much sooner. It failed because this strategy ignored an aim that the game designers and OpenAI thought would be obvious: to finish ahead of others. But no one had told the machine learning system to do this, and it focused instead on what it was told to do—maximize the score—and hit as many targets as it could. To the machine, the strategy resulted in success as it had been told to define success. To a human, the strategy revealed a fundamental misunderstanding.[47]

These kinds of failures, often called "specification gaming," are common. By 2021, DeepMind had collected over sixty examples of them.[48] They read like a list of exploited loopholes. One robot that was supposed to slide a block along a table figured out that it was easier to shake the table. A system for playing tic-tac-toe mastered the art of drawing symbols far away from the board, causing its opponent, a computer program, to crash and resulting in victory. A different agent learned to pause its game, *Tetris*, to avoid losing, while yet another system determined that the optimal strategy was to kill itself at the end of one level in a game to avoid defeat at the harder next level. Maybe most striking of all, a system at DeepMind that was tasked with grasping a ball positioned its claws such that it looked like it was doing so, attempting to deceive the human observers tasked with giving feedback on the machine's performance.

In a way, specification gaming might seem funny—until it's not. If a machine learning system misses the point in the real world, the consequences are likely to be more severe than in games. One neural network designed to detect pneumonia in chest X-rays discovered that X-rays from one hospital were more likely than others to exhibit pneumonia because that hospital usually had sicker patients. It learned to look for the X-ray's hospital of origin rather than at the X-ray itself. Another neural network was designed to identify cancerous skin lesions. It trained on a set of images from dermatologists who often used a ruler to measure lesions they thought might be cancerous. The network recognized that the presence of a ruler correlated with the presence of cancer, and, when looking at new images, started checking to see if a ruler was present rather than focusing on the characteristics of the lesion. In both of these cases, alert human operators noticed the failures before the machine learning systems were deployed to diagnose patients, but it is impossible to know how many cases like these have gone undetected and how many more will go undetected in the future.[49]

Specification gaming by machine learning systems is poised to become more problematic. Reinforcement learning algorithms rely deeply on interactions between the agent and its environment; both algorithms and environments will continue to get more complex. This complexity creates greater possibilities for mis-specification of objectives and for shortcuts overlooked by humans but identified by AI, including deceiving humans themselves.[50] As machine learning becomes more powerful, designers will likely ask more of it, increasing the costs of the failures that arise from specification gaming.[51] As ever, it is hard to tame the new fire.

The risk of specification gaming predates the age of AI, however. It arises anywhere there are misaligned incentive structures, including in human society. As economists often note, once authorities identify a metric and set it as an objective to maximize, people disregard other goals.[52] For example, during the British rule of India, the government grew alarmed about the number of venomous cobras in Delhi and began paying bounties for each dead cobra. At first, this led to fewer cobras slithering around the city, but before long it prompted people to breed cobras and then kill them for the bounties. When the government ended the bounty program, people released the remaining cobras, leading to an overall increase in the population. While the idea of this kind of failure is simple enough, it can be hard to avoid in practice.[53] The complexity and unpredictability of machine learning systems make misaligned incentives even harder to spot.

Viewed this way, addressing specification gaming can seem akin to closing loopholes. But it is worth emphasizing that, in every single specification gaming case DeepMind identified, the machine does not see itself as exploiting a loophole or cheating. It simply tries to achieve the objective its human designers set for it, whether that is scoring the most points in *Coast Runners*, moving a block from one place to another, or figuring out which patterns in its training data indicate that a scan shows disease. The machine does not know the intentions behind that objective or what the designers really have in mind. Nor does it know which aspects of its environment are intentional and which are bugs—like the capacity to endlessly regenerate targets by sailing in a loop in *Coast Runners*. It assumes, so to speak, that everything is as it should be, and then tries to find the best way to achieve exactly what it has been told to achieve.[54]

This literalness means that machine learning engineers must specify a robust and proper objective and confirm that the machine acts as intended in pursuit of that objective.[55] Many instances of specification gaming arise because the objectives the creators set for the machine learning system were not precisely aligned with what they wanted the system to do. The creators should have specified that the agent was to score the most points in *Coast Runners* while finishing the race quickly, to move the block without moving the table, and to look at the lesion itself to determine if it was cancerous or not, ignoring anything else in the image. The disconnect between human and machine produced an undesirable outcome.

But specifying better objectives is hard. To begin, the creators of an AI system must know what they want and don't want the system to do, and they must know the environment in which it will operate. Even in a game as simple as *Coast Runners*, this insight is difficult to gain in advance. AI's capacity to develop unfamiliar but brilliant strategies—like AlphaGo's stunning move 37 against Lee Sedol—can also cause embarrassing failures.

Making things more difficult still, it is tempting to devise objectives that are easily measurable and that correlate with the desired outcome but do not cause it. A team of AI researchers noted how easy it is for real-world applications of machine learning to suffer from this problem. For example, if engineers reward a robot tasked with cleaning a room for leaving no visible disorder in its field of vision, the robot might simply turn off its camera. If they reward it for using cleaning supplies like bleach, it may simply dump them down the drain and claim success. The task of specifying what a clean room looks like is much harder than relying on these other proxy measures of cleanliness, which correlate with evidence of humans cleaning but do not necessarily apply to machines.[56]

The persistence of specification gaming should prompt broader reflection about what machine learning can and can't do. It underscores the challenges of getting machine learning systems to act in the way that their designers desire—to surprise us with their innovative successes but not with their failures. Specification gaming suggests that there is sometimes a gap between what we think we are instructing our creations to do and how they are interpreting the instructions. It also suggests that there is a gap between how we perceive the environment in which machine learning systems operate and how those systems perceive the environment; what is a bug to us is sometimes a feature to them.

Most importantly, specification gaming should guide us toward seeing a deeper failure of machine learning systems. It is a kind of failure that is harder to measure and more philosophical in its nature: no matter how capable they become, AI systems might never actually understand what they are doing.

UNDERSTANDING

In 1980, as expert systems rose in prominence, the philosopher John Searle proposed a thought experiment. Suppose, he said, that you are locked in a room and the only way to communicate with the outside world is via slips of paper passed under the doorway. All of the slips of paper you receive are in Chinese, a language with which you have no prior familiarity. Fortunately for you, however, inside your room is an extensive collection of Chinese books, filled with every possible question and its perfect answer. When you receive a slip with a question on it, all you need to do is match the Chinese characters of the question to the proper entry in a book, copy the Chinese characters of the answer down onto the slip of paper, and send it back out. If the books are complete enough, to the outside world it will look as if you understand Chinese.

Searle argued, however, that one fact is stubbornly true, even if you alone are aware of it: you do not actually understand Chinese. You merely have very good books and you are adept at replicating characters by rote from them. No matter how much your replies impress those outside the room, your lack of understanding remains. When you copy your answers to the questions, you are acting not as a thinker but as a scribe.

Searle's thought experiment might suggest something about what AI is and isn't, what it can do and what it never will be able to. A machine might astound us with its capabilities, he implied, but it does so in the way that the slips of paper with Chinese replies impress those outside the room. The machine has merely learned how to find the right answer from among the data it has been given. If we were to imagine ourselves on the inside of the machine—just as we imagine ourselves inside the room in the thought experiment—we would see that the machine's operations are not enough to produce understanding.[57]

The thought experiment was immediately controversial and ignited a rich philosophical debate. Other theorists offered interesting rebuttals. For example, perhaps you in the room do not understand Chinese, but the system—including whoever wrote those awesomely complete books of questions and answers—certainly does, and you are part of the system. Or perhaps understanding is itself just a form of rote transference. Searle and his interlocutors argued with each other in philosophy journals for decades, parsing who understands what in the thought experiment and what different interpretations mean for AI and for theories of mind.

Most AI scientists, however, landed on a more practical response to the thought experiment: Who cares? As one computer pioneer, Edsger Dijkstra, said, the question of whether computers can think "is about as relevant as the question of whether submarines can swim."[58] Many AI researchers believe that, provided that an AI system does what it is supposed to do, it does not matter if it is mimicking intelligence or *is* intelligence. In their view, if the AI system is capable of better strategic planning than humans, it is irrelevant that the AI does not know the stakes of the battle. Similarly, humans will still be grateful if, say, an AI someday solves a thorny science question and helps develop a new medication as a result. We do not need—and perhaps do not want—the machine to understand what a life is or what it means to save one.

From this perspective, once a system has been proven to work well enough, a lack of understanding is not a failure of machine learning systems in the way that biased performance or specification gaming might be. The key, many AI researchers argue, is to focus on developing more capable machine learning systems, eradicating the failures that do have a demonstrable impact on performance, and leaving the questions of understanding to the philosophers.[59]

Some AI researchers, however, are willing to address the possibility of understanding more directly and to offer a view that implicitly contrasts with Searle's. Ilya Sutskever, the cocreator of AlexNet and the chief scientist at OpenAI, is one of them. He noted that even humans don't really know what it means to understand something, and that it is hard to determine what systems understand. He suggested that the word prediction technology that underpins GPT-2 and GPT-3 could offer some way of measuring how well a system grasps what is happening. "If you train a system which predicts the next word well enough, then it ought to understand. If it doesn't predict it well enough, its understanding will be incomplete," he said.[60] In response, Searle might argue that, if we could see inside even a machine capable of perfect word prediction, it would be the digital equivalent of a room filled with books.

Sutskever remains an optimist about what future AI systems might do. He said, "Researchers can't disallow the possibility that we will reach understanding when the neural net gets as big as the brain," which has more than a hundred trillion connections between its neurons. GPT-3, in contrast, has 175 billion parameters, vast by today's standards but small in comparison to human minds. One of the most significant limiting factors to constructing a larger network, as the previous chapter described, is the computing power required to train it; if compute continues to grow at its current rate or otherwise becomes more efficient, multitrillion parameter networks are perhaps achievable by 2024. Whether they will be capable of understanding is uncertain, and surely depends on one's definition.[61]

The importance placed on machine understanding shapes the kinds of applications for which AI is appropriate and the kind of research that needs to be done before AI can or should be deployed. Although Searle devised his thought experiment in the age of expert systems and not machine learning, we can still learn from the Chinese room idea. If attaining understanding matters, then there are almost certainly circumstances in which it is inappropriate to use a machine learning system. If understanding does not matter, then the focus should be on overcoming other kinds of failures.

It may depend on the stakes. People can sometimes quickly come to trust even those AI systems that do not understand the world in the way that Searle defines the term. No system today comes close to this kind of understanding, and yet drivers use Google Maps, Wall Street traders use high-frequency trading algorithms, and scientists of all stripes use machine learning to aid their inquiries and inventions. In this view, AI is sufficient as a tool; it need not have a mind or an understanding of its own.

Warriors often apply this view to geopolitics. They argue that an AI need not understand what it means to kill in order to target air strikes with more accuracy than humans, and it need not know the difference between truth and lies for it to sow (or fight) disinformation. On the other hand, as Cassandras often believe, maybe AI in geopolitics should be treated differently from AI in mapping applications, stock trades, and science. Perhaps understanding is not the optional final objective but a waystation offering something vital, something without which nations should not go further in deploying their AI.

All of these failures raise another important question: Which AI systems should be invented? The triumphs of Fei-Fei Li, Ian Goodfellow, DeepMind, OpenAI, and others made unmistakably clear the power—and, at times, the frailty—of machine learning. The world of national security took notice. While the evangelists wanted to use AI to advance science and benefit humanity, and the Cassandras worried about machine learning's faults, the warriors perceived a different set of imperatives and pursued a very different set of inventions, adding fuel to the new fire.

FUEL

5

INVENTING

Everyone called it "the Gadget." Some of the greatest minds in the world had helped turn it from only a flicker of an idea to a deadly device. Leo Szilard, a physicist born in Hungary, had conceived of the first possibilities while walking in the London humidity. Years later, he shared his ideas in a letter, signed by Albert Einstein, to President Franklin Roosevelt. After Roosevelt gave the green light, part of the design was conceptualized by Edward Teller, one of the most renowned physicists of the day, with the production overseen by Canadian American physicist Robert Christy. Another part of the design came from John von Neumann, a famed Hungarian mathematician. The Italian Nobel Prize winner Enrico Fermi calculated the Gadget's expected effect. Harvard professor Kenneth Bainbridge was in charge of the test that would determine if Fermi was right, and Robert Oppenheimer, a former Berkeley professor, oversaw the whole project.

From their far corners of the world, these people and so many others had all done their part to invent the Gadget. Now the time had come to see what it could do.[1]

On July 16, 1945, the Gadget stood ready atop a 100-foot tower at the Trinity test site 200 miles from Los Alamos, New Mexico. Many of its creators gathered with their colleagues about twenty miles northwest of the tower. Some had started a betting pool on how big the blast would be.

Fermi went further, taking bets on whether the atmosphere would ignite, and if it would destroy just New Mexico or the entire planet. Some of the soldiers on guard duty found this wager troubling.[2]

At 5:29 a.m. and 21 seconds, local time, the Gadget imploded. The blast tore through the New Mexico sky, mushrooming out across the desert. It was bright as could be, a blinding light that was yellow at first, then red, and then a brilliant purple. The flash was visible in Amarillo, Texas, almost 300 miles east. Eventually, white fog and smoke dominated the scene.[3] The blast was equivalent to 20,000 tons of dynamite.[4] The Gadget was a success.

Bainbridge, the test boss, turned to Oppenheimer, the laboratory director. "Now we are all sons of bitches," he said.[5] Oppenheimer was more poetic; he later said the moment brought to mind the line from the Hindu spiritual book the *Bhagavad Gita*, "Now I am become Death, the destroyer of worlds."[6] Oppenheimer kept a copy of the 700-verse scripture close to his desk. Its teaching of dharma, or duty, resonated with him, though he never could shake the sense that he had unleashed a great terror on the world.[7]

The mushroom cloud over Los Alamos and the ones that followed over Hiroshima and Nagasaki were testaments to human ingenuity and collaboration, even as they were also catalysts for incredible death and destruction. The Manhattan Project that invented the bomb had brought together physicists, engineers, generals, academics, pilots, assemblers, mechanics, guards, photographers, and many more besides, drawn from all sectors of North American and European life. The US Department of Energy, in its account of the test, blandly but succinctly observed:

Despite numerous obstacles, the United States was able to combine the forces of science, government, academia, the military, and industry into an organization that took nuclear physics from the laboratory and on to the battlefield with a weapon of awesome destructive capability, making clear the importance of basic scientific research to national defense.[8]

This kind of collaboration characterized American invention at the time. It was international, drawing on refugees and immigrants to solve thorny scientific and engineering problems. It was vast, pulling contributors from across American society. And it was crosscutting, uniting the worlds of national security, academia, and the private sector; the government led, but others joined. Without such a model of joint invention, the first nuclear weapons might not have been American.

The invention of today's AI is different. The private sector is far ahead in the United States; the federal government has scarcely appeared in the last four chapters. In fact, active distrust of the US government predominates in some technology circles, while reluctance to work with the cumbersome machinery of bureaucracy prevails in others. The specter of a more efficient and capable China emerges as a contrast, causing alarm in Washington that autocracies may pull ahead of democracies in AI. While this overtaking is not inevitable, the future inventions in the field will be determined by those on both sides of the Pacific with grand visions of what the technology can do for science, security, and prosperity.

SCIENCE IN PEACE AND WAR

Of all the notable people who observed the Trinity nuclear test, perhaps the most significant was a man who by 1945 had excelled in academia, business, and government. As the head of the wartime Office of Scientific Research and Development, he was responsible for overseeing 6,000 scientists, including those who worked on the Manhattan Project. His name was Vannevar Bush, and he both embodied and stood at the forefront of American scientific invention during the twentieth century.

Bush had taken a circuitous route in life before finding himself in the New Mexico desert that July morning. He grew up in Massachusetts, where he acquired, in the words of one journalist, a "hard-boiled Yankee shrewdness" and "stern New England features."[9] He earned bachelor's and master's degrees at Tufts by 1913. For his master's thesis, he invented (and later patented) a device that helped surveyors trace the profile of the land they were examining. Many more inventions and patents would follow.

Bush bounced around after graduation. He tested equipment for General Electric, then went back to Tufts to teach mathematics, and then worked for the Brooklyn Navy Yard as an electrical inspector. Next, he returned to school for his PhD, first at Clark University and then in a joint program at MIT and Harvard, where he studied electrical engineering. He completed his PhD in just one year, motivated in substantial part by the desire to earn enough money to get married, which he did after graduation.[10]

From there, Bush hopscotched through industry, academia, and the world of national security. He worked again at Tufts, for a company in the radio business, and then for the American Research and Development Corporation, where he tried to build a device to help the US Navy detect German submarines during World War I. After the war, he returned to MIT as a professor, where he taught courses in electrical engineering. He also worked with several profitable companies tied to his improvements of radios and thermostats.[11] Along the way, he supervised several remarkable graduate students, most notably Claude Shannon, who went on to develop information theory, a bedrock of computer science.[12] All of this was preparation for what was to come.

In 1938, Bush became president of the Carnegie Institution of Washington and also a member of the National Advisory Committee for Aeronautics, the precursor to NASA. It was a comparatively dull time for popular science—the cutting-edge inventions at the World's Fair in 1939 were a Mickey Mouse watch and the Gillette safety razor—but Bush remained steadfast in his belief that advanced engineering had much to contribute to society and security.[13] As World War II loomed and then raged in Europe, he remembered how, during World War I, civilian scientists and military leaders had been at odds with one another. To prevent that from happening again, he proposed a federal agency to improve the coordination of scientific research and development in the service of national security goals. In June 1940, he met with President Roosevelt. After only fifteen minutes of conversation, Roosevelt signed off on his idea.[14]

Bush became chairman of the hastily created National Defense Research Committee. He appointed two leading military officers and two leading civilian scientists to the committee and they got to work, with Bush managing the group's projects. One of the effort's most significant early successes came with the refinement of radar technology, achieved in significant part at MIT. Improved radar eventually enabled Allied planes and antiaircraft guns to better anticipate and thwart German bombing raids over Europe, helping to turn the war in the Allies' favor. Engineers at MIT's Radiation Laboratory later summed up the prevailing view about the power of their invention: "The atomic bomb may have ended the war, but radar won it."[15]

By June 1941, Roosevelt had seen enough from Bush to make him the head of the new Office of Scientific Research and Development. It was a big step up and a testament to the belief that science, whether in the form of radar or medical discoveries such as penicillin, contributed mightily to the war effort. Bush reported directly to the president, and Congress endowed the office with an ample budget and a broad mandate to invent technologies that would benefit American national security. By this point, Bush was doing far more managing than engineering. He found himself overseeing 2,500 contracts worth more than half a billion dollars (equivalent to around $9 billion in 2022); the United States invested around $3 billion in 2022's money in radar alone. Bush was responsible for research in subjects as disparate as missile accuracy, the effectiveness of antibiotics, and—most secret of all—atomic weapons.[16] It was this last one that brought Bush to New Mexico, and led him to quietly tip his cap to Oppenheimer when the Gadget imploded.

In July 1945, as the war was ending and the nuclear age was dawning, Bush published his reflections in a widely read essay. He praised the spirit of collaboration that enabled the American military machine, writing, "It has been a war in which all have had a part. The scientists, burying their old professional competition in the demand of a common cause, have shared greatly and learned much. It has been exhilarating to work in effective partnership."[17]

In his essay, Bush did not ignore the dangers of arms races. He knew that science could be deployed in the service of destruction; he would later argue unsuccessfully that the United States should not test the more powerful hydrogen bomb and should instead pursue arms control talks with the Soviet Union. To fend off the possibility that science could be used for ill, Bush proposed that humans could invent machines to store and transmit information, making knowledge more accessible and promoting greater collaboration among nations. In a separate report to President Truman—titled *Science: The Endless Frontier*—Bush left no doubt that he saw the federal government as central to such an endeavor. He advocated for robust federal government investments in fundamental research, arguing that it was essential both for the national economy and for national security.[18]

Bush thought that science could strengthen democracy and democracy could strengthen science. "I believe that the technological future is far less dreadful and frightening than many of us have been led to believe, and that the hopeful aspects of modern applied science outweigh by a heavy margin its threat to our civilization," he wrote. "[T]he democratic process is itself an asset with which, if we can find the enthusiasm and the skill to use it and the faith to make it strong, we can build a world in which all men can live in prosperity and peace."[19] In the face of heavy obstacles, he favored consequential government action over hand-wringing and doubts.

Though Bush's influence in Washington waned as he aged, the muscular role of the federal government in science remained in place, especially when it came to national security. Bush became known as the visionary who had established this partnership.[20] "No American had greater influence in the growth of science and technology than Vannevar Bush," said Jerome Wiesner, President Kennedy's science advisor. His most enduring legacy was a model of collaboration that relied not on large government laboratories but instead on contracts with universities and leading industrial researchers.[21] Indeed, these contracts were significant. By 1953, the Department of Defense dedicated $1.6 billion per year to scientific research (just under $16 billion in 2022's dollars), and 98 percent of the money spent on physics research in the United States came either from the Pentagon or from the Manhattan Project's successor, the Atomic Energy Commission.[22]

The impact was felt beyond physics. In 1965, the Department of Defense funded 23 percent of all university science research in the United States. Electrical engineering and eventually computer science and AI became vital to national security. Labs at wealthy corporations, such as Bell Labs and Shockley Semiconductor Laboratory, also propelled science forward, sometimes with the aid of government funds. For example, government contracts for major corporate laboratories responded to the need to shrink transistors so thousands of them could fit in a single computing device, as required by the missile and space programs. In the 1960s, NASA was responsible for purchasing upward of 60 percent of integrated circuits, the successors to the early transistors.[23] Eventually, Department of Defense money would lead to the creation of the internet's precursors, while the National Science Foundation and other agencies would fund Google's cofounders' tuition when they were doctoral students at Stanford. It is difficult to overstate the role of government funding in incubating the information technology revolution.

Not everyone supported this joint model of American invention, and some computer scientists actively rejected it. Perhaps the most famous among them is Geoffrey Hinton, the descendant of a storied British scientific family. Hinton is the great-great-grandson of George Boole, who invented the binary logic used by computers; the great-great-grandson of the noted surgeon James Hinton; and the great-grandson of the mathematician Charles Howard Hinton, who conceptualized much of four-dimensional geometry. Hinton's father was a foremost expert on insects, and his father's cousin was a researcher at the Los Alamos laboratory. Hinton felt the pressure of this scientific success his entire life. His mother told him that there were two paths open to him: "Be an academic or be a failure."[24] His father, without irony, repeatedly told Hinton, "Work really hard and maybe when you're twice as old as me, you'll be half as good."[25]

When Hinton was a teenager, a friend showed him a hologram and explained how it worked, with many different beams of light bouncing off an object and the information about them stored diffusely.[26] Hinton had an intuition that the human brain worked the same way, holding and processing information across its vast network of neurons. While a student at Cambridge University, he focused on trying to understand and mimic how humans thought. But as he completed his undergraduate studies in 1970, he realized that he had not made much progress. Worse still, neither had anyone else. With seemingly little hope of understanding the mind, he decided to become a carpenter; given his mother's earlier admonition, one can imagine her displeasure.

After a year in carpentry, Hinton decided to give academia another try. He studied at the University of Edinburgh, the oldest AI research center in the United Kingdom. He won a spot with a well-known PhD supervisor who at first shared his intuitions about how AI might work. After Hinton was admitted but before he arrived, however, his supervisor changed his mind, concluding that neural networks would never work. Hinton stuck to his guns anyway, even though some of his weekly meetings with the supervisor ended in shouting matches.[27]

Hinton nonetheless stuck it out and he received his PhD in AI in 1978. With few job prospects in Britain, he moved to the University of California in San Diego to become a postdoctoral fellow alongside a group of cognitive psychologists who were interested in his ideas about how computers and humans could think similarly. Together, they made progress on developing ways for neural networks to update their parameters as part of a learning process, the very foundations of the machine learning paradigm.[28] Neural networks were almost universally dismissed at the time and had been out of favor for almost a decade, but Hinton continued researching the subject.[29] "People said, 'This is crazy. Why are you wasting your time on this stuff? It's already shown to be nonsense,'" he later recalled.[30]

Hinton eventually moved to Carnegie Mellon as a professor. He stayed for almost six years before deciding to leave the United States, where, as a socialist, he had grown disillusioned with the Reagan administration's defense policies. He especially disliked the administration's aggressive foreign policy in Latin America, a region from which Hinton and his wife had adopted two children. Perhaps most significantly, he resented the fact that the Department of Defense—and not a civilian organization— was a primary funder of AI research in the country (Hinton nonetheless had accepted nearly $400,000 from the Office of Naval Research in 1986 to fund some of his AI work).[31]

Another democracy offered an alternative. Canada, eager to develop its technology sector, had a government program to fund unorthodox ideas; at the time, neural networks certainly qualified as such. In 1987, the Canadians lured Hinton to Toronto with an opportunity to conduct further fundamental research on neural networks. He eventually became a computer science professor at the University of Toronto despite never having taken a computer science course himself.[32]

Hinton and neural networks both remained on the fringes of academia as the rigid rules of expert systems dominated AI research. For decades, he stayed in Toronto, supervising the few PhD students and postdoctoral fellows who had the same unusual faith in machine learning as he did. Among them were Yann LeCun, one of the other pioneers in neural networks, and Ilya Sutskever and Alex Krizhevsky, who created AlexNet under his supervision. "I was one of the very few who believed in this approach," Hinton said, "and all the students who spontaneously believed in that approach too came and worked with me. I got to pick from the very best people who had good judgment." Looking back, Sutskever recalled, "We were outsiders, but we also felt like we had a rare insight, like we were special."[33]

During the 1990s and 2000s, Hinton thus found himself in an unusual position: he was a professor of legendary scientific lineage who had become a contrarian both in AI and in American defense policy. He was one of the undisputed leaders of an academic field nearly everybody thought was useless, self-exiled to a city far from Silicon Valley and his native Britain. Undeterred, he plowed ahead.

After Ng's 2012 experiment at Google X showed the capability of neural networks backed with massive computing power, it became clear that Hinton was in fact correct: neural networks could accomplish what expert systems simply could not. Hinton, who was sixty-five years old at the time and had suffered from depression for much of his life, felt the pressure of his family's towering scientific history lift a little. "I finally made it, and it's a huge relief," he said.[34] He thought there was a lesson in his long path to success and the decades he spent on the sidelines pursuing an idea very few people on the planet thought had any potential. "In the long run, curiosity-driven research just works better," he said. "Real breakthroughs come from people focusing on what they're excited about."[35] Seemingly implicit in his critique was his skepticism about short-term corporate and military projects driven by the imperatives of profit or security.

With the meteoric rise of neural networks that followed Ng's success, Hinton became one of the most recognized names in AI and helped make Canada a hotbed of AI research. He has continued to do pathbreaking work, inventing new designs for neural networks.[36] Others followed in his footsteps, pushing the technological frontier forward and leading to the developments described in the first three chapters. Even more notably, at least for national security, some other pioneers came to share Hinton's suspicion of government funding and the military applications of AI. Demis Hassabis, the cofounder of DeepMind, is a prime example.

In 2014, Hassabis and his team were ready to sell DeepMind to Google. The AI company was ascendant, bringing many of the best machine learning scientists and engineers on Earth together in London. It had made key innovations in reinforcement learning, and future break-throughs lay ahead. Google was ready to pay more than $600 million to buy the firm, but it wasn't just about the money for Hassabis. The Deep-Mind team negotiated another condition into the terms of sale: that nothing that the company built could ever be used for national security projects. "We have committed to not ever working on any military or surveillance applications, no matter what," Hassabis said, distilling one of the core beliefs of many evangelists. AI was a tool to be used for the good of all.[37]

The sale of DeepMind with that condition attached was a watershed moment, at least in retrospect. During the nuclear age and the early era of computing, governments were the main attractors of talent, including in AI; today, academia and private sector labs are enlivening the machine learning moment. In the United States, nearly every single modern machine learning breakthrough has emerged from nongovernmental sources. The tide has definitively shifted.

Many of the most brilliant AI thinkers, like Hassabis and Hinton, explicitly reject the notion that they have an obligation to work on issues of national security. In a democracy, no law can reasonably force them to do so. The US government must win their trust or go it alone. In 2017, as AI became a more urgent national security concern, the warriors at the Pentagon sought to bring AI evangelists into the fold. To do this, they turned to one of the military's most adept generals.

PROJECT MAVEN

The first movie that young Jack Shanahan ever saw on the big screen was about the Battle of Britain in World War II. The movie had been filmed partly at the airfield in Duxford, England, where the Shanahan family temporarily lived while his father, a professor of mathematics, was on sabbatical at Cambridge University. Shanahan immediately found the aerial dogfights between Allied and Axis pilots captivating. He knew then that he wanted to fly.

After college, Shanahan joined the US Air Force. He became a weapons systems officer, flying in the F-4 and then the F-15 fighter aircraft. As he rose through the ranks, he delved into other subjects, including intelligence and reconnaissance; Shanahan is fond of saying that he "became a jack of all trades and a master of none."[38]

Perhaps his most grueling and eye-opening assignment was as the US military's deputy director for global operations from 2011 to 2013. The anodyne title obscures what Shanahan called "a brutal job." In the span of just two years, he dealt with the earthquake, tsunami, and nuclear disaster in Japan, the aftershocks of the Arab Spring, NATO's interventions in Libya, and a near war with Iran. His purview included information operations, electronic warfare, and special programs—"the things I really can't get into," he said.[39]

Some of these secret programs worked and some didn't. With others, he still doesn't know if they made a difference. But from his years spent overseeing such activity, he came to appreciate more deeply the value of collecting intelligence and of subtly nudging global affairs in a direction more favorable to the United States. For as much as the military was prepared for combat, he said, sometimes "maybe a bomb was not the best answer for something you were trying to do." When he left Global Operations for his next assignment, one of his going-away presents was a chart from Operation Bodyguard, the World War II deception strategy that helped enable the D-Day invasion. It was a reminder of the importance of the work carried out in the shadows.[40]

In 2015, Shanahan became the director for defense intelligence for warfighter support in the Pentagon. Part of his responsibility was to improve the processing, exploitation, and dissemination of information collected by American spy planes, drones, and other surveillance tools. "It was a problem that was just unmanageable," he said. "Far too much data coming off of platforms and sensors—higher-quality data than had ever been put off these sensors before. Multidomain data was coming in at unprecedented volume, variety, usually of high veracity. And we couldn't deal with it." The Pentagon needed to invent a way to make sense of all of this information.[41]

Shanahan's team wanted something revolutionary. They found that most military projects attempting to tackle the problem were then in their early stages, at best. For inspiration, the team turned to a group of advisors known as the Defense Innovation Board. Created in 2016, the organization was charged with bringing the technological innovations of Silicon Valley into the Pentagon. Its chairman was Eric Schmidt, the former CEO of Google.

Shanahan told Schmidt how much information the military was collecting from its various cameras and other sensors and how hard it was to make sense of everything. Schmidt laughed. Shanahan recalled him saying, "'What do you think YouTube takes in every day? Yeah, your problem is big, but it's not exactly unmanageable. Go figure this out in commercial industry.'"[42] It was a more polite version of what Schmidt had in 2016 told Raymond Thomas, the four-star general who at the time led the United States Special Operations Command: "You absolutely suck at machine learning. If I got under your tent for a day, I could solve most of your problems."[43]

To both generals, Schmidt argued that, unlike in the past, this time the military wasn't alone in facing its challenges. In the 1940s, the creation of the atomic bomb was a unique project, but in the 2010s the need to analyze data was ubiquitous. Shanahan didn't have to invent everything from scratch. He could find some of what he needed in the private sector, especially with machine learning.[44]

To test what machine learning could do, Shanahan and his team decided to analyze the videos collected by the cameras on American unmanned aerial vehicles. These drones surveilled hot spots all over the world, sending back information for large teams of intelligence analysts and military units to examine. It was often a boring and thankless job. Shanahan's group put together a pilot project that used supervised learning to try to automate at least some of it. They first contracted with a company to label the objects that appeared in old videos to serve as training data, then developed an algorithm that could learn from those labels and recognize similar objects in footage unanalyzed by humans. It was like AlexNet for the battlefield, spotting tanks, jeeps, and more.

In April 2017, Shanahan showed a demo to then-deputy secretary of defense Robert Work, who had long harbored big hopes for how AI could help the Pentagon. As soon as Work heard about the project, he loved it. "'That's it, now I've finally found my team. You are henceforth called

the Algorithmic Warfare Cross-Functional Team,'" Shanahan remembered Work telling him. It was the kind of bureaucratic-ese name that
would appear on official documents but that few people would actually
use in practice. From the start, everyone called Shanahan's effort Project
Maven.[45]

Even though it started with analyzing intelligence, Project Maven was
the Pentagon's first major foray into using AI for much more. The leadership believed that a fight of algorithm against algorithm would be likely
one day—especially if the United States and China went to war—and
that such a battle would require the military to make faster and more
intelligent moves than its adversaries. Work was clear about the overarching purpose of transforming warfare. "'Jack,'" Shanahan recalled him
saying, "'we're standing this up because you're the only ones that I see
doing this, but it's not about just intel. We've got to move beyond intel.'"
The Defense Innovation Board was blunter still: "'We're going to lose to
China someday if you do not embrace this idea called artificial intelligence and machine learning,'" he remembers its members telling him.[46]

Shanahan needed to scale up his pilot project quickly. He looked for
more private-sector partners. Even though the researchers at DeepMind
were off the table, Google remained enticing. As Schmidt had implied,
the company knew how to collect, store, and analyze data perhaps better than any organization on Earth. As a leader in machine learning that
employed thousands of top computer scientists and engineers, Google
was well-positioned to improve the automated analysis at the core of
Project Maven. Throughout the summer of 2017, Shanahan and his team
built a relationship with the firm's executives in California.

One of Google's leaders at the time was Fei-Fei Li, the creator of
ImageNet, who was on sabbatical from Stanford. As the company was
getting ready to sign the contract with the Pentagon for Project Maven,
Li wrote an internal email—later leaked to the *New York Times*—that outlined both her hopes and concerns. "It's so exciting that we're close to
getting MAVEN!" she wrote, and suggested that the company do "a good
PR story" about the value of working with the Pentagon. She offered one
warning, however: "Avoid at ALL COSTS any mention or implication of
AI. Google is already battling with privacy issues when it comes to AI and
data. I don't know what would happen if the media starts picking up a

theme that Google is secretly building AI weapons or AI technologies to enable weapons for the Defense industry."[47]

Shanahan later said that he advocated a different approach, one that was more transparent about what the company was doing for the Pentagon, but Google executives rejected his views. He said that they chose to downplay the role of AI in Project Maven not just in the media but also internally. "'Let us handle it,'" Shanahan recalled them saying. "'This is very sensitive right now. We just don't want to put this out in the company yet.'" Shanahan, eager to get to work, deferred to them on the matter. In his eyes, Project Maven fit easily into the United States' rich history of collaborative invention on matters of national priority. It soon became clear that not everyone agreed.[48]

DISSENT

If someone had told Meredith Whittaker in 2006 that she would be at the center of a pitched battle between the world's most powerful corporation and the world's most formidable military, she would have blanched at the thought. Back then, she was focused on more prosaic tasks—namely, looking for work. She posted her résumé on Monster.com, the precursor to LinkedIn, where Google took note of her.[49]

Whittaker had studied English and rhetoric at Berkeley, so she wasn't scouted for a top-tier software engineer job or anything close to it. Instead, she started as a temporary worker doing technical documentation and customer support. It was her job to write up what Google's engineers were doing and to help people use the products. But her talent was obvious. She moved up to working on technical standards and then later founded a project within Google known as the Open Research Group, which increased collaboration with technologists working outside the company.

During Whittaker's early years at Google, the big technology policy debate of the moment was focused not on AI but on net neutrality, the question of whether internet service providers could charge more for certain kinds of data or whether they had to treat all internet traffic the same. Whittaker, like many in the technology world, held the latter view. She worried that differential pricing would undermine the open nature of the internet and give an advantage to large corporations. To prove the case, in 2008, she helped devise a project called M-Lab to collect data on traffic as it moved about the internet.

As an undergrad, Whittaker had read critical theory—the critique of entrenched power structures—but M-Lab was her first real education in politics, or, as she put it, "power in action." In particular, the experience drove home for her the importance of bolstering rhetoric with data, and it revealed how companies and governments spin the available information to suit their positions. "The key question was, how can you tell if the net is neutral?" she said. "This seemingly simple provocation ended up taking up seven years of my life." The experience taught her the importance of how data is collected and who dictates how it is used. Technology and politics are inextricably intertwined.[50]

In the course of her years at Google, she gained a reputation for speaking up when she thought the company was making a mistake. Colleagues would often come to her for advice when they felt the same. Whittaker, who was not primarily working on AI at the time, said she learned about Project Maven in the fall of 2017, when fellow Googlers indicated they were uncomfortable with it. Whittaker realized that AI was the frontier where modern power was in action. She started to explore further.[51]

As she learned more about Project Maven, Whittaker was shocked to find that Google was in talks with the Pentagon. She worried that the company's machine learning systems would eventually be used to target military drone strikes that she, like many human rights activists, believed were illegal under international humanitarian law. More generally, she feared what a partnership between the Pentagon and the tech giant would portend. "It was the wedding of an extremely powerful multinational corporation (that has access to extraordinarily intimate data about almost everyone) with the world's most lethal military . . . without any democratic debate or deliberation," she said.[52]

Many at the company shared her view. After Google announced that it had won the Project Maven contract in September 2017, concerns grew as more employees discovered the role that AI would play in the effort. In addition to Whittaker, other Googlers, including at DeepMind and in the cloud computing division, raised objections internally. The response from the company was muted, so the employees started posting to internal message boards, explaining their worries in more detail. The posts quickly attracted attention and support, and the dissent started to snowball.

Emboldened by the internal response, Whittaker and other concerned colleagues drafted a letter to Sundar Pichai, the CEO of Google. They wrote:

> We believe that Google should not be in the business of war. Therefore we ask that Project Maven be cancelled, and that Google draft, publicize and enforce a clear policy stating that neither Google nor its contractors will ever build warfare technology. . . . We cannot outsource the moral responsibility of our technologies to third parties.[53]

Within a day, the letter accumulated a thousand employee signatures. Google's leadership called an all-hands meeting in response. At the meeting, Diane Greene, the head of Google Cloud, tried to quell the rebellion. She told employees that Project Maven was not for offensive purposes and that Google was not aiming to build technology for warfare. The discussion that followed indicated that internal concerns remained strong. In one employee's account, "One woman stood up and said something like, 'Hey, I left the Defense Department so I wouldn't have to work on this kind of stuff. What kind of voice do we have besides this Q&A to explain why this project is not okay?'"[54] The question fit within the evangelist worldview charted by Hinton and Hassabis: AI needed to be kept as far away as possible from geopolitical competition and combat.

Sergey Brin, the cofounder of Google, responded, noting that Google was unusual in that it permitted such dissent and questioning, and that many companies would not have tolerated the months of turmoil from employees about a relatively small contract. Even some of the people most upset about Project Maven agreed with this argument; Google had been built from the start to be less hierarchical than other technology firms, though the company still made mistakes and failed to think through the implications of its products. But in the moment and to many employees, Brin's comments came off to many employees as dismissive and out of touch. Within a day or two, more than a thousand additional employees signed the letter to Pichai.

One employee said that "the letter spread like wildfire" after the all-hands meeting and that the campaign to kill Google's involvement in Project Maven "began to take on a life of its own."[55] Some employees resigned over the issue.[56] Others tried to ensure that questions about Maven came up at every all-hands meeting; still others talked to reporters to garner media coverage. Some of the dissent took on a classic Google form when employees used an internal program called Memegen to make "funny" and "dark" memes about the company leadership's decision to keep the contract.[57] In a very Silicon Valley move, executives reportedly tracked the spread of these memes to measure the strength of the opposition.

The dissent put the changing face of American innovation into stark historical relief. In the 1930s and 1940s, a collection of global talent joined together to build the atomic bomb; in 2018, a different assembly of talent from all over the planet united at Google to resist Project Maven. One organizer of the campaign said that the role of employees overseas was vital and that "many of the Googlers who supported the campaign come from regions of the world where the American military has been extremely destructive."[58] To some of these employees, the argument that Google had to assist the American military because Pentagon officials said the country's security was at stake seemed dubious at best and duplicitous at worst. Google, in their view, was an international firm with billions of users nearly everywhere; the United States had the world's most powerful armed forces already and—no matter what the country's leadership said about China—was not under any serious threat. In this view, there was no need to build tools for the Pentagon. Whittaker argued that Maven was "effectively making Google beholden to the interests of the US military."[59]

On April 4, 2018, the internal dissent burst out into mainstream view. The *New York Times* published the employees' letter and revealed that it had more than 3,000 signatures. While the dispute had already earned coverage in the technology press, this was the first time it made its way into more widely read publications. Google once again tried to put the issue to rest, repeating many of its previous statements about Project Maven being "non-offensive" and arguing that "the technology is used to flag images for human review and is intended to save lives and save people from having to do highly tedious work."[60]

The spin didn't help. A week later, *Defense One* reported that Project Maven was actually a pilot program for a much broader collaboration between Google and the military. In particular, Google sought to bid on a $10 billion contract, the Joint Enterprise Defense Initiative, or JEDI, which would entail hosting a large amount of the Pentagon's information in Google's cloud servers.[61] To Whittaker and others, this report revealed the company's duplicity; Google had reassured employees that Maven was a small project and nothing more, but the employees came to think that, in actuality, Google aimed to use Maven as a foothold to climb into the ranks of major defense contractors. One employee said, "Exposing all of those lies damaged leadership on this issue just as much as anything else that we did. That loss of trust really hurt them."[62]

The opposition only grew. Employees at other technology companies signed a letter calling on Google to withdraw from Project Maven.[63] In May, a letter from scholars studying technology and war added still greater support to the campaign, drawing notable signatories, including Terry Winograd, a colleague of Li's at Stanford and the PhD advisor of Larry Page, one of Google's cofounders.[64] The company was under fire from all sides, and there was only one way forward.

On June 1, Google backed down. It announced that it would complete its obligations under the Project Maven contract it had signed for 2018, but that it would not seek the renewal of the contract beyond that time.[65] Pichai pledged that Google would not develop AI for "weapons or other technologies whose principal purpose or implementation is to cause or directly facilitate injury to people."[66] Shanahan later said that the company told him "they had no choice." He said he was disappointed in Google's decision. "There was fault on all sides," he later acknowledged, "But I do place the lion's share of the blame back in the company's lap" for failing to be transparent about the contract.[67] Shanahan didn't say it, but others in the world of national security found it richly ironic that Google—a company that had distanced itself from its old "Don't Be Evil" motto, had been the subject of a substantial number of privacy investigations over the years, and had reportedly worked with the Chinese government on censoring search results—was trying to claim the moral high ground by refusing to continue work on Maven.[68]

For Whittaker and her colleagues—some of whom were critical of other Google projects—the dissent on Maven was about principle as well as process. Whittaker argued that, even if Google had been more transparent, the company's participation in Project Maven was "inherently problematic." It would enable what she saw as "US militarism or imperialism" and would do so in a way that was disproportionately likely to hurt people in poverty, people of color, and people thousands of miles away from Google's headquarters in sunny California. She feared that the programs would be used to enable lethal operations conducted with little oversight and visibility, and that Googlers would not even know what was done with the tools that they built. By inventing these machine learning capabilities for military use, they were surrendering their leverage and enabling a system with the wrong incentives. "The fundamental question is about power and accountability," she said.[69]

Shanahan thought accountability was possible. In 2018, he became the first head of the Pentagon's newly established Joint AI Center, which aimed to centralize and expand the military's use of AI. He assisted with the development of the Pentagon's AI principles. These were core concepts that aimed to govern how the military deployed the technology and to alleviate the concerns of Cassandras like Whittaker.[70] For Shanahan, the principles highlighted a difference between how democracies and autocracies deployed AI. He wanted to show that democracies "do care about things like ethics that our adversaries don't necessarily care about."[71] Whittaker remained unpersuaded. She found the Pentagon's principles vague and unenforceable. Like many technology company employees, she still did not feel a need to build advanced technologies for the US government. "Ethical principles without accountability are marketing," she said.[72]

Others felt differently. Some firms, including Microsoft and Amazon, continued to work with the Pentagon, mollified in part by these principles. Microsoft and Amazon competed fiercely for the $10 billion JEDI contract that Google had coveted. An array of traditional defense contractors and smaller technology firms took over Project Maven from Google; they likely did not have the top-tier talent that the search giant did, but they helped Shanahan move the project forward. To work with these companies, Shanahan and others tried to streamline the Pentagon's convoluted procedures, painfully slow procurement processes, and rigid budgeting culture. It was a sign that the gap between Silicon Valley and the Pentagon was not just about worldview—though those divisions were real—but also about more mundane matters.

In their own way, both Shanahan and Whittaker welcomed the debate about ethics, or at least appreciated it in retrospect. Whittaker and her colleagues wanted to draw an ethical line and show that technology employees could refuse to work on projects they thought were dubious. In Silicon Valley, where top talent is extremely scarce, especially in AI, it is the employees, not the executives, who sometimes have the most leverage. Shanahan wanted to stress the urgency of improving the Pentagon's contracting and oversight processes and show that collaboration was possible, even if not everyone agreed. It was better, in his view, to determine the relationship between the public and private sectors at a moment of comparative peace rather than when a crisis loomed and time was short.[73]

In 2019, Whittaker resigned from Google. She said that she had faced retaliation for her activism over Project Maven and for organizing employee walkouts, which she led after the company made a $90 million payout to a top executive accused of sexual misconduct. After leaving Google, Whittaker moved full-time to the AI Now Institute at New York University, which she had helped found in 2016 to research "the social implications of AI systems."[74] As one of the most prominent voices in the world of technology advocacy, she had come a long way from job hunting on Monster.com.

In 2020, Shanahan retired, closing out a thirty-six-year military career. In an interview a few months before he left government service, he recalled a quote from one of his early mentors in the military: "No friction, no traction." Project Maven had certainly provided its fair share of friction, but the general seemed optimistic nonetheless. He remained confident that it was urgent and vital for the US military to put AI to work, and that Maven had helped orient the Pentagon in the right direction. "You make progress in an organization by fighting it out in the crucible of pressure and disagreement," Shanahan argued, not by "saying we all agree with each other."[75]

But while the United States fights it out at home, another country seems more unified in its mode of invention.

INVENTING AI FOR AUTOCRACY

The Association for the Advancement of Artificial Intelligence promotes the development and responsible use of AI. In addition to supporting numerous academic publications, every year it hosts one of the field's largest and most significant research conferences. In 2016, as the annual gathering concluded, the organization announced the date for the next meeting: January 2017 in New Orleans.

Immediately, trouble emerged. The proposed dates conflicted with the Chinese New Year, meaning that many Chinese researchers would be unable to attend. After a period of reconsideration, the conference's leadership recognized that these researchers were now such a force to be reckoned with in the AI community that the conference needed their work. The organizers changed the dates to hold the event a week later.[76]

The decision underscored a broader shift in AI research. That year, researchers in China had almost as many papers accepted through the conference's rigorous peer review process as researchers in the United States. It was a result that had seemed unthinkable just a few years before, when China lagged far behind, and inevitable just a few years later, when the "citation impact"—a shorthand measure of academic influence—of Chinese publications grew rapidly. One study concluded that "China's rapid growth in AI research output surpassed that of the EU and the United States in 2015 and 2018, respectively."[77]

Even when Chinese research was comparatively rare at leading academic conferences on AI, one of the exceptions was Tang Xiao'ou. He was born in China but moved to the United States to pursue his master's degree and PhD. He earned the latter in 1996 from MIT, where he focused on computer vision, teaching machines to recognize objects in images. At the time, the task seemed nearly impossible. Tang was undeterred and committed to the problem for the long haul. "It took me a couple of years to find out what I really liked to work on at MIT. Then it took more than 20 years of hard work in this area such that we finally start[ed] to see some real industrial applications and impacts. It does not happen overnight," he said.[78]

Tang did most of this work as a professor at the Chinese University of Hong Kong, with the exception of a sabbatical at Microsoft Research Asia. Along the way, he met a number of other researchers focused on problems in computer vision. Together, they deployed machine learning to dramatically improve image processing. Tang and his team started developing a facial recognition system that they called DeepID. In 2014, after several years of improving the system's algorithms, feeding it more data, and increasing the size of the neural network and the amount of computing power it used for training, Tang and his team had developed a version of DeepID that was ready to take on some major challenges.[79]

Just as the ImageNet database provided a way to evaluate how image recognition systems recognized objects in pictures, a data set called Labeled Faces in the Wild offered a mechanism for judging facial recognition systems. In 2014, the state-of-the-art facial recognition system was known as DeepFace, invented by a team of Facebook AI researchers; the company had access to massive data sets and a huge incentive to solve the facial recognition problem, given the importance of photos on its platform. In its test, Facebook's DeepFace recognized more than 97 percent of faces in the Labeled Faces in the Wild data set correctly, slightly worse than human performance. By 2014, Tang's DeepID could do even better.[80] It was one of the most notable breakthroughs ever for a group of Chinese AI researchers.

Tang's work was primarily academic in its focus, though the real-world applications were apparent to some, including a venture capitalist named Justin Niu. In October 2014, Niu stopped by Tang's office for a demonstration of the technology. Tang showed Niu a scene from Tiananmen Square, perhaps the most famous and infamous destination in China. It was the spot where Mao Zedong had declared the founding of the People's Republic in 1949 and the location of his mausoleum. Most notably, the square was where Chinese security forces repressed protests demanding greater freedoms in 1989, part of a broader crackdown in which they killed hundreds of people. Decades later, the Chinese government censors any mention of the massacre and even many references to its date, June 4.[81]

In hindsight, this history provides important geopolitical subtext for what Tang showed Niu. In his demonstration, heavy smog obscured the square, partly hiding many of the people from view. Tang showed how DeepID could pick out the parts of the scene that were most important, ignoring the obstructions and focusing on the faces in the image. At the time, just two years after the first ImageNet competition and just months after DeepID had triumphed over Facebook's system, such a capability was remarkable. Niu was shocked by what Tang and his colleagues had invented. He and other venture capitalists invested $30 million before Tang had time to finish filing the paperwork for the new company.[82]

For the company's name, Tang used a phonetic transcription of the Shang dynasty and its first ruler, also named Tang; the firm is often rendered in English as SenseTime. The Shang dynasty began in 1600 BCE, a period of Chinese ascendance. To Tang, this reference was deliberate. "China was leading the world then," he said, "And in the future, we will lead again with technological innovations." He wanted SenseTime to be a part of that renaissance.[83]

Tang and his cofounders spent much of 2015 transitioning from academic research to establishing SenseTime as a business. In 2016, they began unveiling commercial products. Their first big hit was a mechanism for verifying peer-to-peer payments. These payments were a giant part of the Chinese economy, which lacked a consumer banking industry as developed as the one in the United States. SenseTime's facial recognition software ensured that the person holding the phone and sending the payment was in fact the authorized user on the account.[84]

SenseTime's technology only got better from there. The company began producing a string of remarkable results that vaulted the field of computer vision forward. In 2016, it won three components of the ImageNet competition, with its machine learning systems proving most adept at detecting objects, parsing video, and analyzing what happened in scenes. The next year, SenseTime had forty-three papers accepted at the two leading computer vision conferences, beating American technology giants Facebook and Google.[85]

The financial world took notice. In 2017, SenseTime took in an additional $410 million in venture capital investment, the largest second-round investment ever for an AI company. In April 2018, it raised another $600 million in capital to fund its continued research and development.[86] It put this money to work, acquiring top AI talent to build better algorithms, storing data collected from its rapidly growing customer base, and assembling a massive amount of computing power—including more than fifty-four million GPU cores.[87] By the end of 2018, just over four years after its creation in an academic lab, SenseTime was valued by some at $7.7 billion, making it the top AI startup in the world.[88]

The company enjoyed a truly global reach in both business and research. Major international brands, like the computer chip maker Qualcomm and the car manufacturer Honda, raced to partner with the firm.[89]

The Chinese phone company Xiaomi enlisted SenseTime to improve its software for generating photo albums. TikTok deployed its algorithms to make its users look better in their videos.[90] Cities from Shanghai to Singapore trumpeted strategic alliances with SenseTime and highlighted new lab openings.[91] Major universities in America, Asia, and Australia set up partnerships. Some of these involved vast collaborations; for example, the company funded twenty-seven projects for AI researchers at MIT.[92]

In many respects, China was both the ideal incubator and marketplace for a startup like SenseTime. "China is really moving ahead, especially in video and image understanding, because we have the real-world problems, we have the real-world data, and we also have a stronger talent pool dedicated to those kinds of things," the firm's CEO, Xu Li, said.[93] SenseTime also benefited from the fierce competition offered by China's other AI champions, and from the country's massive scale. The leader of SenseTime's augmented reality business left Microsoft to return to Asia after a conversation with a friend. In the United States, she said, "We were struggling to get a thousand users; then I talked with my friend who was working at a startup in China, and she said, 'Oh, a million users is nothing—we get that in several days.'"[94]

Not all of SenseTime's customers were interested in phone payment apps and video filters, however. Tang spoke of a desire to work with government, saying in 2018, "More efforts are needed to strengthen the partnership between State-owned and private companies when exploring the application of AI."[95] Indeed, some of the company's most notable business was in security, selling facial recognition software to analyze video from China's vast network of surveillance systems. The company leads China's national working group on facial recognition standards and its leadership has acknowledged that the government is the firm's largest data source.[96] China's Ministry of Public Security, which operates more than 200 million cameras in the country, uses the company's technology via third-party resellers. The company's software bolstered some of the cameras' systems to better track and analyze people as they go about their daily life. In 2019, this technology accounted for 35 percent of the firm's revenue.[97] If there is another massacre at Tiananmen Square, there seems to be little doubt that a camera running SenseTime's software will be there to capture it.

In 2018, Leon, a Chinese policing company, formed a joint venture with SenseTime to aid a crackdown against the Uyghur Muslims in Xinjiang that the US government and other observers call a genocide.[98] SenseTime's software—alongside the software of other major AI companies—helped the police and security services track individuals' movements and recognize their faces. The surveillance enabled by the technology is nearly ever-present, and the consequences of falling under suspicion are severe. Chinese authorities currently detain more than a million Uyghurs in what the government euphemistically calls "re-education camps."[99] Perhaps fearing international backlash, SenseTime ended its formal role in the Leon joint venture in the spring of 2019.[100]

In part because of SenseTime's other government ties, the backlash followed nonetheless. In October 2019, the Trump administration placed SenseTime and other major Chinese firms on the Entity List, which made it harder for American companies to sell their products to SenseTime. As justification, the administration said that the company was among those "implicated in human rights violations and abuses in the implementation of China's campaign of repression, mass arbitrary detention, and high-technology surveillance against Uyghurs, Kazakhs, and other members of Muslim minority groups."[101] This sanction caused universities like MIT to review their own partnerships with the company, though the university did not return donations.[102] Most importantly, though, the move cut SenseTime off from its supply of American-made computer chips, a critical resource for deploying more complex and capable machine learning systems. In response, the company's leaders expressed their disappointment with the decision and publicly mused about trying to build their own computer chips to become more independent from the American supply chain. Doing so would tie the company more tightly to Chinese vendors and align it still further with Chinese government interests.[103]

SenseTime is not the only company in the Chinese government's orbit, as the government is gaining leverage throughout the country's private sector. The government has sought to create a "dual-use technology and industrial base," embodied in a strategy known as military-civil fusion.[104] This strategy seeks to harness the combined weight of state-backed enterprises, Chinese manufacturing giants, private equity, and universities affiliated with the People's Liberation Army.[105] Chinese President Xi Jinping elevated the importance of military-civil fusion in 2016 and established a Central Commission for Integrated Military and Civilian Development to drive the application of new technologies, logistics, and training for defense purposes. In addition, Xi oversaw the passage of two laws, one related to cybersecurity and one to intelligence, that gave the government further tools to compel information and gain assistance from corporations and citizens within China.[106] Since then, he has only expanded his efforts to push China's private sector to serve the needs of the state, making clear that even powerful companies are not wholly removed from the obligation to advance government interests.[107]

While SenseTime helps the Chinese government tighten its grip at home, other major AI companies directly aid the Chinese government's foreign intelligence analysis efforts. The opportunity to do so arises in part because Chinese government hackers have conducted a decades-long cyber espionage campaign that, among other successes, has gathered information on hundreds of millions of Americans. This campaign has breached major companies—including leaders in the credit reporting, travel, and healthcare sectors—as well as government agencies, including the US Office of Personnel Management, which holds tens of millions of security clearance records on government employees, contractors, and their family members. Contained within the vast data sets obtained by hacking is enough information to find undercover CIA officers, economic espionage targets, and other opportunities for intelligence activities.

American spies have observed extensive data transfers between Chinese companies and the Chinese government and concluded that the government compels or pressures these firms to use AI to find the secrets within hacked data sets. "Chinese technology companies play a key role in processing this bulk data and making it useful for China's intelligence services," said William Evanina, the top counterintelligence official in the United States when the data transfers were uncovered.[108]

The United States does not have the same influence over its technology sector. While it can compel data on foreign targets under the Foreign Intelligence Surveillance Act, it cannot task major firms with answering analytic questions. An intelligence official envisioned to *Foreign Policy* what an advantage such help would be:

Just imagine on any given day, if NSA and CIA are collecting information, say, on the [Chinese military], and we could bring back 7, 8, 10, 15 petabytes of data, give it to Google or Amazon or Microsoft, and say, "Hey, [examine] this on the weekend. We want all these analytics; get it back to us next week."[109]

As the two nations press onward, each in its own way and each with a wary eye on the other, one point is obvious: just as world leaders recognized in the 1940s that the next age would be atomic, policymakers in both the United States and China believe that AI is at the core of their competition. Long before lethal autonomous weapons fight the wars of the future, more mundane matters like government relations with technology giants will determine who has power and how it is wielded. The AI inventions each side produces will be key to warfare and geopolitical competition—and both the United States and China are trying to invent the future first.

6

KILLING

"We're gonna blast them now! We will die, but we will sink them all—we will not become the shame of the fleet."[1] Soviet naval officer Valentin Savitsky uttered these fateful words aboard a B-59 diesel-powered submarine on October 27, 1962, at the height of the Cuban Missile Crisis. As US President John F. Kennedy and Soviet leader Nikita Khrushchev brought the world to the brink of war above the water's surface, another showdown was unfolding in its depths.

The situation was dire. US Navy destroyers were tracking Savitsky's submarine near Cuba and had dropped explosive charges as a signal to the crew to surface and surrender. Instead of risking capture, the submarine dove deeper to evade attack and, as a consequence, lost all incoming radio traffic with Moscow. Savitsky and his men didn't know whether war had broken out, only that the attacks against them were becoming more frequent and intense. The batteries were running low, and carbon dioxide levels were rising. The air conditioning had failed, causing the temperature to reach 141 degrees Fahrenheit in the diesel room.[2] As the circumstances worsened, Savitsky turned to the one major arrow left in his quiver: the submarine's nuclear torpedo.

Commodore Vasily Arkhipov, the man in charge of the entire Soviet submarine flotilla, was also on the submarine. Arkhipov, calm and intelligent, had a legendary reputation and had survived a nuclear submarine meltdown the previous year. He was no stranger to superstition; his wife recounted that he once burned all of their love letters to ward off any bad luck during his deployment.[3] Fearing imminent death, shorn of communications links with the Kremlin, and with time running out, Savitsky told Arkhipov that World War III had begun, moving to launch the nuclear torpedo.[4] But Arkhipov, by rank and reputation, had earned a vote in the decision to use nuclear weapons.

Arkhipov voted no. He reasoned with Savitsky, extolled the value of patience, and suggested they wait for more information. In the end, the crew never fired their weapon. By averting a cascading atomic conflict, Arkhipov's actions may have quite literally saved the world.[5]

Sixty years later, this episode from the annals of the Cold War seems remote, but it raises fundamental questions about the role of human judgment in the use of force. Even at the height of the Cuban Missile Crisis, when humanity peered into the abyss, political and military leaders—not machines—made the choices that determined the fate of millions. Whether one laments Savitsky's impulsiveness or praises Arkhipov's patience, there remains no doubt that humans were in control.

Thanks to the rapid advances in AI, however, the human role in warfare is changing. Weapons systems with ever more autonomous functions will speed up decision-making. They will amplify the precision of ordnance fired at adversaries, allow humans to focus on strategic decisions, and enable military technologies and operations that were not possible before. These include unmanned submarines, drone swarm warfare, and self-targeting missiles.[6]

Major governments know that the new flames of war are afoot. After nearly two decades of counterinsurgency conflict in Afghanistan and Iraq, the United States has repositioned itself for an era of strategic competition with China and Russia. The US 2018 National Defense Strategy notes that AI, among other new technologies, will be critical to ensure the country can "fight and win the wars of the future."[7] Russia and China are likewise investing in AI in pursuit of technological superiority on the battlefield, though they do not always have the same strategic objectives as the United States.[8]

Machine learning can transform a range of military functions—including routine but vital back-office and logistics tasks, as well as intelligence analysis—but automating violence is the subject that poses the starkest strategic and ethical questions.[9] By most reasonable definitions, lethal autonomous weapons—the focus of this chapter—are hardly hypothetical. Some estimates find that governments already spend around $11 billion on such weapons annually, with billions more spent on other kinds of autonomous military capabilities, such as surveillance and reconnaissance drones.[10] This investment, coupled with the rapid technological progress of AI, increases the urgency of the debate around their development and use.

The Cassandras—and many evangelists—weigh the risks and debate the moral and legal implications. Advocates, scholars, and researchers from The Campaign to Stop Killer Robots and The International Committee for Robot Arms Control argue that all countries should forswear lethal autonomous weapons. In their view, such weapons are killing machines incapable of human reasoning and lacking the judgment necessary to make truly informed decisions about the use of force. By delegating crucial decisions to such flawed machines, governments not only risk violating their commitments under international humanitarian law but also surrender meaningful control over the use of force and offend basic moral sensibilities. In this view, the only appropriate way forward is to ensure that every weapon on the battlefield is under human control.[11]

The warriors contend that it is wrong *not* to integrate autonomous functionalities into weapons systems. Senior civilian and military officers believe that lethal autonomous weapons can help them safeguard national security. They note that the Cassandras often presume the United States will be the aggressor in any conflict, but that the primary benefits of autonomy may be in the areas of defense.[12] By forfeiting the initiative and forgoing potential strategic benefits, democracies invite defeat and risk squandering everything in a naive grasp at idealism. As the pace of warfare accelerates, warriors imagine a future in which commanders issue the orders and denote appropriate targets, but machines execute the missions, select among targets, and absorb the risks. Warriors believe that progress toward development of such weapons will protect troops in the field, ensure more accurate operations, and produce fewer civilian casualties and less collateral damage.[13] Democracies, in this view, must be ready to win a new kind of war.

ALGORITHMIC WARFARE

After twenty-seven years of service in the US Marine Corps, Robert Work had seen enough of warfare to know when change was coming. As the deputy secretary of defense who created Project Maven and placed General Jack Shanahan at its head, Work recognized the transformative potential of AI on the battlefield. "The next major revolution will roll out of what I call algorithmic warfare," he said. His is perhaps one of the clearest and most consistent voices arguing that combat will change from a fundamentally human endeavor to one in which machines interact at speeds beyond the capacities of human cognition.[14]

History informs Work's view. Prior to the United States' combat operations in the first Gulf War in 1991, warfare mostly relied on unguided munitions launched in massive salvos. This approach was by necessity, since the bombs weren't very accurate. During World War II, the so-called circular error probable—the radius of the circle around a target in which a bomb would fall half the time—was around 1,000 meters, or just over 3,000 feet. It was in many cases impossible to aim with accuracy at specific targets, and planners would sometimes evaluate success by studying how many acres of a city a bombing raid had destroyed.[15]

The first Gulf War, however, revealed how much combat had changed. Facing off against Iraq, the United States showcased the fruits of a technological revolution that had been several decades in the making. The technology of modern conflict included improved intelligence collection capabilities, better battlefield communication mechanisms, more accurate weapons, and stealthier systems. These sophisticated capabilities were the product of what is sometimes called the Second Offset, a strategy through which the United States built cutting-edge military technology as a means of thwarting the Soviets' larger military forces. "Net-centric warfare," as commanders called it, put the Second Offset into practice, linking together these various high-tech components to improve control and precision in combat while denying key information to enemies.[16] Today, net-centric warfare enables precision-guided munitions to consistently hit within three meters of their target, many thousands of times more accurate than their World War II counterparts.[17]

The United States' decisive victory against Iraq in 1991 proved to other nations that they couldn't match the American technological advantage.[18] "The only competitor that could have hung with the United States was the Soviet Union, and it disappeared," Work said. "China was way behind in high-tech warfare in the 1990s, so the United States knew it would have enormous freedom of action" in deploying military forces to accomplish its objectives. Joining information superiority and lethality became a hallmark of the American way of war.[19] While it didn't always translate to success in later counterinsurgency operations in Afghanistan and Iraq, this concept was at the core of the United States' strategy to prepare for and deter conventional conflict with China and Russia.

The Pentagon knew, however, that its technological advantage would not last forever. Andrew Marshall, known as "the Pentagon's Yoda" for his strategic insights, repeatedly warned Work and others of this fact.[20] China and Russia would soon catch up to the American innovations of the past. To keep its advantage, the United States would have to change the game again.

Work believed that Marshall was right. By the time he took office in 2014, China was trailblazing new means of negating the American technological edge. Instead of matching the United States bomber for bomber or carrier group for carrier group, China aimed to attack the logistics, communications, and satellite networks that enabled the United States to fight far from its shores.[21] The goal wasn't to beat the United States in conventional battle but to blind American information systems via electronic warfare and other means. Without a technological advantage, the United States' military commanders would struggle to project power in the Pacific.[22]

In large part due to Work's efforts, the Pentagon adopted what it called the Third Offset strategy. "The theory," Work said, "is to make our networks more effective than Chinese battle networks." He wanted the various components of the US military to work together and interfere with the other side's ability to do the same. Whereas the Second Offset introduced the idea of modern precision-guided weapons and improved battlefield communication, the Third Offset took it further, using AI-enabled systems and human-machine collaborations to speed up decision-making and foster a faster tempo of operations.[23] The more autonomy weapons have, the less reliant they are on vulnerable communication links back to American commanders and the more quickly they can act in combat.[24]

Work came to see the battle network as a broad enterprise for managing the use of force amid the uncertainty and chaos of war. In his vision, commanders would set priorities at the highest levels, as manifested in policy guidance and military orders. They then would delegate some decisions to an authorized entity in the battle network, such as an autonomous aircraft. This authorized entity would be empowered to take actions, including firing weapons, within specific boundaries. If a commander decided to attack an enemy tank position, for example, he or she would designate the tanks as a target and then delegate to a machine the choice of the best course of action to engage those tanks. This system for managing combat, Work argued, would ensure that there are "appropriate levels of judgment over the use of force"—the standard stipulated by Department of Defense policy guidance for the use of AI.[25] Commanders would determine the goals, but the system would be flexible and fast enough to win modern wars and thus hopefully deter conflict.[26]

Work emphasized that in this arrangement, humans would remain at the core of the battle network even as the technology got better. The lethal autonomous weapons would take their direction from American commanders, and those commanders would be held accountable if they mistakenly deployed the wrong weapon or failed to remediate a malfunctioning one. The US military would not develop weapons that operated entirely on their own, setting their own strategic priorities and choosing their own missions, even if it was technologically possible to do so. "What commander would ever utilize a weapon like this?" Work asked. To do so would be to surrender one's control over the battlefield.[27]

As an example of what would be feasible and appropriate in his vision of the battle network, Work pointed to the Assault Breaker II program at DARPA, the research and development agency at the Pentagon. The original Assault Breaker was a Second Offset capability from the Cold War built to thwart a potential massive Soviet armored assault in Europe. The newer version could potentially be used to defeat Russian tanks or to target Chinese aircraft or bases. In a moment of crisis, American bombers would launch Assault Breaker II missiles. The missiles would fly to an area of concentrated enemy forces and hover. Each missile would release smaller munitions, and each of these would select and attack an enemy target.

In DARPA's vision, commanders would delegate authority to Assault Breaker II to select individual targets, but humans would be responsible for the overall decision to deploy the weapon and the general area toward which it would go.[28] "So let's say we fire a weapon at 150 nautical miles because our off-board sensors say a Russian battalion tactical group is operating in this area," Work hypothesized. "We don't know exactly what of the battalion tactical group this weapon will kill, but we know that we're engaging an area where there are hostiles." It would be up to the machines to fly autonomously to the designated zone, determine the best targets based on predesignated criteria, and attack those targets while minimizing collateral damage.[29]

This capability is only one part of the future of war as Work and other planners see it. A sophisticated enemy would also likely try to jam communications links, which would require machines to be able to operate independently. To prepare for this possibility, in 2015, DARPA started a program called CODE—short for Collaborative Operations in Denied Environment. It was one of the most significant forays into building the battle network of the future; the program manager said that CODE's drones would learn to hunt together, "just as wolves hunt in coordinated packs with minimal communication."[30]

Within a few years, the CODE program showed promise. One February morning in 2019, a swarm of six TigerShark drones took off over the desert in Yuma, Arizona. The TigerSharks were some of the newest drones in the US Air Force's arsenal, designed to be much cheaper, smaller, and more versatile than some of their more famous predecessors, like the Predator and the Reaper. The six TigerSharks worked together to perform a variety of tasks independent of human direction or control. As the exercise planners simulated various scenarios, the swarm responded to the changing tactical situations. Whereas other unmanned aerial vehicles would have reacted to a loss of communications with human commanders by executing their previously programmed mission, the CODE software enabled the drones to "autonomously share information and collaborate to adapt and respond to different targets or threats as they pop up," according to DARPA.[31] After the announcement of the successful test, the Pentagon said the US Navy would take over the program and develop it further.[32]

Another DARPA project, AlphaDogfight, showed that algorithms could defeat even the best human pilots, at least in simulations.[33] In August 2020, eight AI systems battled it out in virtual fighter jet duels, with the winner—built by a small company called Heron Systems—advancing to a final round against a military pilot in an F-16 simulator. Heron's algorithm easily defeated its human opponent, five to zero. It was an exemplar in miniature of many of the key features of AI, including reinforcement learning and self-play; Heron estimated that its system had learned through four billion simulated dogfights. At the end of the day, the conclusion seemed similar to what had happened with chess, Go, and *StarCraft* before it: traditional tactics had failed. "The standard things that we do as fighter pilots aren't working," the defeated human said.[34]

Yet another US military test in late 2020 offered a deeper sense of what autonomy across battle networks could do. In the Arizona desert, the Army conducted an exercise designed to simulate combat in 2035. Autonomous ground, air, and space systems conducted reconnaissance, then transmitted their findings to another autonomous system that analyzed the information for signs of enemy targets. A different automated system recommended which weapons to employ against the identified targets. Human operators signed off on the system's proposed method of attack, and then the autonomous systems carried out the strike.

The Army said the network of autonomous systems showed substantial improvements in speed and capability. "This is happening faster than any human could execute. It is absolutely an amazing technology," said Ross Coffman, the general in charge of overseeing key parts of the Army's future warfare systems. The tests showed that a process that often takes twenty minutes—identifying the target, choosing a weapon, coordinating with other units, and firing—took only twenty seconds when machines did most of the work.[35] Other tests showed that frequent human interruptions hindered the automated systems more than they helped; in one set of simulations, soldiers working with teams of ground-based robots and aerial drones performed worse when the humans micromanaged the autonomous systems and did better when they gave the machines more autonomy.[36] Another test, by DARPA, had similar results. "Actually, the systems can do better from not having someone intervene," a DARPA official said.[37]

This continued improvement of machines in combat hints at the varying strategic and ethical complexities of autonomous weapons.[38] The tests also lead to an obvious concern, especially for Cassandras: if humans are already getting in the way, competitive pressures will dictate that before long they will see their roles diminish still further. For now, at least, the US military rejects the notion that such a shift is appropriate or inevitable. "Obviously the technology exists to remove the human . . . but the United States Army, an ethical based organization, [is] not going to remove a human from the loop to make decisions of life or death on the battlefield," Coffman said.[39] Whether autocracies—or anybody—will constrain themselves when a war is raging is at best uncertain.

So far, many of these tests and debates have played out with land- and air-based systems, but the battle network extends under the ocean too. When he was deputy secretary of defense, Work unveiled an autonomous submarine known as the Sea Hunter. They were far cheaper than the Navy's multibillion-dollar ships and could be produced in large quantities. Work elaborated on the possibilities:

You can imagine anti-submarine warfare pickets, you can imagine anti-submarine warfare wolfpacks, you can imagine mine warfare flotillas, you can imagine distributive anti-surface warfare surface action groups. We might be able to put a six pack or a four pack of missiles on them. Now imagine 50 of these distributed and operating together under the hands of a flotilla commander, and this is really something.[40]

Work's implication was striking, and a sign of how far things had progressed from a few decades before: in the next submarine crisis between superpowers, there might not be a Savitsky or an Arkhipov aboard—in fact, there might not be any humans at all. But what would an automated Arkhipov do in a crisis? In answering this question, one researcher spoke for many Cassandras when she feared the worst.

THE ONLY WINNING MOVE

Lucy Suchman has made it her mission to understand and improve the relationship between humans and machines. As a doctoral student in anthropology at UC Berkeley in the 1980s, Suchman became fascinated by how humans construct knowledge about the social world and then use this knowledge to create technical systems, from the photocopier to computer graphics and animation to modern-day AI and robotics. It was fitting that she ended up at PARC, the Palo Alto Research Center at Xerox that had refined the most essential means of interaction between humans and computers: the mouse.

Suchman's early work focused on the preprogrammed rules of expert systems. She marveled at how "extremely impoverished" the systems were in their knowledge about the social world. These primitive attempts at AI needed everything to be measured and quantified before they could have any sense of what was happening. "Machines are completely reliant on state changes," she observed. If an environmental factor was not captured in the inputs to the machine's rules, to the machine it was like it did not exist.[41]

Her research raised fundamental concerns about the limits of machines' ability to gain situational awareness and make appropriate decisions. In the 1980s, when President Ronald Reagan launched the Strategic Defense Initiative, the space-based missile defense system known as the Star Wars program, she grew increasingly concerned about the automation of military systems. In 1983, she helped form a group called Computer Professionals for Social Responsibility.[42] The group advocated against what she described as "highly destabilizing and inherently unreliable" launch-on-warning systems—mechanisms that would enable the United States to retaliate against a Soviet nuclear strike while enemy missiles were still in the air. She thought that such a strategy risked starting a war if computer sensors erroneously suggested that an attack was under way, or if the software miscalculated.[43]

As Suchman learned more about the application of AI to weapons systems, she came to appreciate the importance of shaping policy debates as an activist. She became involved with the International Committee for Robot Arms Control and later joined the discussions surrounding the United Nations Convention on Certain Conventional Weapons in Geneva.[44] It was organizations like this, she thought, that could highlight some of the ways in which machines could transform warfare for the worse.

For Suchman, the first crucial issue in lethal autonomy was target identification. She was unconvinced that the ground- and space-based missile defense systems that were at the core of Reagan's ambitions would be able to detect a nuclear attack without false positives. "There was substantial evidence that it was extremely difficult to do that in a reliable way," she said. "There were numerous examples where those systems got confused by moon rises and other sorts of unexpected configurations of the world." To have a consequence as serious as nuclear war depend on rigid rules and flawed sensors was to court catastrophic failure.[45]

In the age of machine learning, the automation of target selection and the initiation of force is more problematic than ever, Suchman argued. Neural networks may offer more flexibility and insight than their expert system predecessors, but this is still not enough. The contingencies and uncertainties of war require strategic intuition and the ability to make judgments about legitimate and nonlegitimate targets based on context— qualities that Suchman believes machine learning systems do not have. As she concluded in a 2016 paper on the topic, "human level discrimination with adequate common-sense reasoning for situational awareness would appear to be computationally intractable."[46]

Even the rapid progress of AI in some key areas did not sway her from this conviction. In her view, an image recognition system, such as AlexNet or the ones Project Maven at the Pentagon eventually built, may be able to identify what it is looking at, but it will struggle to grasp the context required to determine what is a legitimate target. "That is the aspect of human-computer interaction and of artificial intelligence that has continued to resist much progress," she said. "Automation only works in domains where you've really got everything fully specified. You don't take a domain that's already fraught with uncertainty and automate it."[47] The chaos of conflict zones, in her view, made the automation of war particularly unwise.

Suchman's second major concern was that the likely means for performing and improving target verification—adding more training data— would only serve to exacerbate and obscure biases. Presuming that lethal autonomous weapons would be trained on past decisions to use force, she pointed to what she called the "fatal flaws" in the United States' drone program. She believed that poor human decision-making, in part due to stereotypes and confirmation bias, led to drone strikes that killed civilians.[48] Using such decisions as training data for future machine judgments about what would constitute an imminent threat would invite the same kind of failures described in chapter 4. The machines and the processes were simply not as good as many warriors believed.

Taken to their logical conclusion, Suchman's concerns about biased training data opened her up to a counterargument often made by proponents of what AI can do: machine decision-making might never be perfect, but it could still be better than the flawed human decisions of the status quo. History offers cases of machines getting it right while humans got it disastrously wrong. For example, in 1988, the crew of the USS *Vincennes* mistook Iran Air flight 655 for an attacking F-14 Tomcat and shot it down, killing all 290 people aboard. The ship's onboard sensors had correctly concluded that the plane in question was ascending as a civilian airliner would, not diving to attack; a later investigation concluded that the crew either misinterpreted the data or concluded from other sources that they were under attack and disregarded the ship's information.[49]

Cassandras like Suchman have a notable rejoinder: just because humans make catastrophic errors does not necessarily mean that machines are more capable of managing modern combat. "My approach is: show me," Suchman said. "If somebody comes along and could convincingly show me a system that had greater sensitivity to the nuance of life on the ground in a relevant area of military operations, I would be open to that."[50] Not every critic of lethal autonomous weapons agrees with her position—some believe that such weapons are illegal and immoral under any hypothetical circumstance—but Suchman is willing to stake a key portion of her argument on machine learning's capabilities, or lack thereof.

Other cases bolster Suchman's response and show that machines sometimes get it wrong too. During the US invasion of Iraq in 2003, commanders deployed the Patriot missile system to guard against Iraqi missiles. The Patriot system—which did not use machine learning but did use earlier forms of automation to present recommendations to human operators—shot down all nine missiles the Iraqis fired during the war, with American troops authorizing each decision to fire. Unfortunately, however, the human-machine teams that operated the Patriots also shot down two friendly aircraft. In one instance, a Patriot's software mistook an allied plane coming in for a landing as an enemy missile designed to destroy American radar. Even after the military put additional precautions in place, another Patriot identified an enemy ballistic missile where none existed. In response to the Patriot's warning about the missile, the human commanders switched the machine to a mode in which it could fire on its own, at which point it immediately launched two missiles, both of which hit an American plane in the area and killed its pilot.[51]

In addition to concerns about a machine's capacity to grasp context and avoid bias, Suchman argued that the lack of transparency in modern AI posed a major problem for any machine learning system involved in killing. The identification and selection of human targets "are delegated to the system in ways that preclude deliberative and accountable human intervention," she said. Even if there was oversight of lethal autonomous weapons, such systems would have no way of explaining to human overseers why they were recommending a particular course of action. The "meaningful human control"—Suchman's preferred alternative to the Pentagon's "appropriate levels of human judgment"—would be minimal, as humans would have little capacity to interrogate the machine's decision-making process.[52] Instead, as another critic of autonomous weapons put it, any human oversight would be reduced to simply determining whether to press the "I Believe" button and accept the machine's choice without the ability to understand it.[53]

Finally, Suchman argued that, while the laws of war do not explicitly restrict autonomous weapons, the case for banning them rests on how these systems operate in practice. She and many other critics of lethal autonomous weapons point to the Martens Clause of the preamble to the 1899 Hague Convention, which states that "populations and belligerents remain under the protection and empire of the principles of international law, as they result from the usages established between civilized nations, from the laws of humanity and the requirements of public conscience." Killing with autonomous weapons systems, Suchman argued, deprives the targets of their dignity, renders accountability impossible, and falls short of the standards of public conscience. It is a subject of robust and largely unresolved debate, especially since the Martens Clause has never been used to ban particular kinds of weapons and because it is not clear what the public conscience demands, if anything, on the question of lethal autonomous weapons.[54]

Setting aside the debate about the Martens Clause, in order to satisfy the laws of war any lethal autonomous weapon must accord with the rules of distinction, proportionality, and precaution to attack. These rules already govern, at least in theory, human and state behavior in warfare. They stipulate that wartime activities distinguish between combatants and civilians and minimize excessive incidental damage to civilian targets. When attacks do not meet these standards, under the laws of war they are not acceptable.[55] The degree to which a lethal autonomous weapon would be capable of following these rules is debatable—hence Suchman's concerns about the inability of autonomous weapons to identify legitimate targets and distinguish noncombatants.

As someone with a deep understanding of the history of interactions between humans and machines, Suchman acknowledges the allure of lethal autonomous weapons. She simply doubts that it will be possible to realize such ambitions in an effective, ethical, and legal way. The rhetoric and ambition of Work and other warriors reminded her of her formative experiences of campaigning against Reagan's Star Wars initiative. In both cases, she advocated against "the fantasy of the perfect, fully integrated ... command and control system." While there is no doubt that AI has made enormous progress since the 1980s, Suchman said she believed that there remains "a gap between [the theory] of net-centric warfare and the realities of how complex distributed operations actually go."[56]

Fog and friction are intrinsic to war, she argued, and autonomy won't resolve the uncertainty they create. Autonomous systems will indeed shrink the time from identifying a threat to selecting and engaging the target, yet that doesn't mean they will be better at the job. In her view, machines will be faster than humans, but just as or even more likely to err as humans. We will simply "exacerbate the problem by speeding everything up," she said. The only way to control weapons in the midst of uncertainty is to ensure that a human keeps a finger on the trigger at all times and is accountable for what the weapon does.[57]

There is ample support for this view among some governments, scientists, and citizens. In November 2013, the United Nations Group of Governmental Experts adopted a mandate to determine what to do about "killer robots." Since then, thirty countries have come out in favor of banning them.[58] More than 4,500 experts in robotics and AI—often evangelists for what the technology can do in other arenas, such as science and medicine—have signed on to calls for a ban on offensive lethal autonomous weapons, as have more than 60 nongovernmental organizations.[59] Leading individuals and organizations in science, including Stephen Hawking, Elon Musk, Demis Hassabis, and the American Academy for the Advancement of Science, have come out strongly against such weapons, with some believing that they could cause a global arms race and Musk saying international competition for AI is the most likely cause of World War III.[60] A survey of citizens in 26 countries found that 61 percent of them oppose such weapons. The UN secretary-general has also called for a ban, and the International Committee of the Red Cross has advocated for significant limits.[61]

As Arkhipov intimated with his warning on that October day in a submarine off the shores of Cuba, sometimes the only winning move is not to play.[62]

THE HIGH GROUND

For Work, not playing is a losing move, both practically and morally. His foundation for this view highlights three questions at the core of the American debate about lethal autonomous weapons. First, how significant is the threat from other nations, especially from China? Second, how essential are lethal autonomous weapons to neutralizing that threat? And third, what are the moral implications of using such weapons?

According to Work, the threat from China is central to the United States' defense priorities. He believes that the constraints imagined by Suchman and others would put the United States at a major disadvantage. A ban, or even aggressive requirements for human oversight of minute details in the heat of battle, invites defeat against adversaries whose weapons and warriors are not similarly encumbered. "Just imagine in a major conflict with China where you have missiles going all over the place and all of your people are following the weapons into the target rather than worrying about what the Chinese are doing," he said. "It's a sure way to get your ass kicked."[63] Such constraints, he suggested, would be akin to ceding the high ground to the enemy and fighting uphill.

There is some evidence to support Work's view that China intends
to pursue lethal autonomous weapons, but it is hard to draw definitive
conclusions about its goals for using them. China's approach, which is
characterized by muddled statements, seems to reflect a desire for pre-
serving options in the future and leaving American policymakers in the
dark about its capabilities. In a December 2016 paper for the UN Group
of Governmental Experts, China indicated support for developing a bind-
ing system to constrain lethal autonomous weapons, but in its position
paper in April 2018, the Chinese delegation muddied the waters further,
urging only consideration of "the applicability of general legal norms."
In the latter paper, China emphasized that it supported only a ban on
the use of such weapons, not their development—a position that seems
duplicitous.[64]

In addition, the 2018 position paper offered a definition of lethal
autonomous weapons that is not shared by others, including Work.
Among its characteristics were that the weapons could not be controlled
or recalled once released and that they were "indiscriminate" in their
killing effect.[65] No current or planned weapon meets that definition. Any
weapon of that sort would obviously violate international law as it currently stands, and there would be little strategic value for commanders
in deploying systems that they could not control at all. By offering such
an outlandish definition of lethal autonomous weapons and by banning only their use and not development, China appears to be giving
itself cover to pursue the creation of slightly more constrained and discriminating weapons systems.[66] Nor is this pursuit hypothetical; American officials allege that China has already begun developing basic lethal
autonomous weapons—such as helicopter drones with machine guns,
with more advanced stealthier drones to come—and exporting them to
other nations.[67]

In Work's view, China will before long claim the advantages of
autonomy even if the United States does not. "As humans ascend to the
higher-level mission command and I've got machines doing more of
that targeting function, those machines are going to be challenged by
machines on the adversary's side and a human can't respond to that.
It's got to be machines responding to machines," he said. "That's one of
the trends of the Third Offset, that machine on machine [combat]."[68] To

Work, this trend seems inevitable, whether democracies like it or not; to guard against China's inventions, the United States will need its own.

Work argued that lethal autonomous weapons offer a moral high ground too. He noted that the revolution in net-centric warfare and guided munitions was an ethical advance because it limited incidental casualties and provided a foundation for still more progress. Human-directed weapons are not by definition more moral. "What about spraying someone with napalm?" he asked, a brutal method of killing that humans have routinely employed. Moving forward with a ban on lethal autonomous weapons, Work contended, could result in a prohibition on every guided munition under development, weapons that have increasing levels of autonomy but also increasing levels of accuracy.[69]

Improvements in autonomous weapons systems, on the other hand, would make war more moral, enabling commanders to better detect and calculate possible collateral damage, employ more targeted force, and protect ships, bases, and ground vehicles from modes of attack that are too fast for humans to process. Even better, reducing the need for humans in combat would lessen the risk of needless death on both sides. "What about targeting tanks and subs and planes? When you do that, you are killing humans. Worse, you are equating humans in a sub as parts of a machine. That robs them of human dignity," Work said.[70] A machine-versus-machine conflict, with careful targeting and both sides' commanders exercising appropriate oversight, might be a better outcome if the goal is the preservation of human life, especially civilian life. "It is a moral imperative to pursue this hypothesis," Work said.[71]

What sets democracies apart from autocracies, Work continued, is the overall framework in which lethal autonomous weapons would be deployed. Every use of such weapons would be under an umbrella of democratically legitimate civilian control of the military that is both pragmatic and principled.[72] More generally, many warriors believe, as the Defense Science Board recommended, that

whether mediated by man or machine, all acts, but especially acts related to warfighting, must be executed in accordance with policy and so, in some sense, there is no completely autonomous behavior. Any use of autonomy must conform to a substantive command and control regime laying out objectives, methods and express limitations to ensure that autonomous behavior meets mission objectives while conforming to policy.[73]

In this view, democracies could defend themselves, reduce the need for bloodshed, and claim the moral high ground all at once.

Work recognized that the prospect of war against a less constrained enemy would prompt tough choices. "We might be going up against a competitor that is more willing to delegate authority to machines than we are, and as that competition unfolds, we'll have to make decisions about how to compete," he said, a theme echoed by other American military planners.[74] When he was still at the Pentagon, Work once raised the question more bluntly: "If our competitors go to Terminators . . . how would we respond?"[75] In such circumstances, Work worries that the United States might focus on the moral issues but that "our potential competitors may not."[76]

He knew such a situation would pose major challenges for the United States. Referring to ever-growing autonomy, Work said, "The only way that we would go down that path, I think, is if it turns out our adversaries do and it turns out that we are at an operational disadvantage because they're operating at machine speed and we're operating at human speeds." The enemy has influence over what form combat will take, and it is hard to know what it will be. "The nature of the competition about how people use AI and autonomy is really going to be something that we cannot control and we cannot totally foresee at this point," he said.[77]

Suchman found such theorizing about threats and the compromises required to meet them deeply speculative and unnecessary. When pressed about the imperative for the United States to win a conflict against an autocracy like China, should one arise, Suchman pushed back. She did not see lethal autonomous weapons as essential to such a victory and did not see the threat as being nearly as imminent or as serious as Work did. On the contrary, she worried that the weapons' failings would make them unreliable in critical moments, kickstart destabilizing arms races, and bring unintended consequences. The world would suffer as a result. "If the hyper-developed states fight deadly wars between themselves," Suchman concluded, "there's always going to be fallout for everybody else."[78]

In some sense, the perceived moral high ground depends on the strategic context. Larry Schuette, the director of research at the Office of Naval Research, compared the debate about lethal autonomous weapons to the debates between American war planners before the surprise attack on Pearl Harbor. What Americans saw as appropriate on December 8, 1941, was much more aggressive than what they had seen as appropriate on December 6 and in the decades before; activities that were unthinkable one day became inevitable the next. "We went all of the twenties, all the thirties, talking about how unrestricted submarine warfare was a bad idea [and that] we would never do it," he said, referring to the practice of sinking an enemy nation's noncombatant merchant ships without warning. "And when the shit hit the fan the first thing we did was begin executing unrestricted submarine warfare."[79]

In fact, views changed even before the calendar did. Just hours after the attack on Pearl Harbor, American submarines received orders to begin unrestricted submarine warfare against Japan. For Schuette, the lesson from history is clear: "We are going to be violently opposed to autonomous robotic hunter-killer systems until we decide we can't live without them."[80] The transition from one view to another could happen as quickly as a spark turns into a fire.

It is not yet the modern equivalent of December 8, 1941. But if one believes that there is a threat from a sophisticated autocracy—a matter of substantial disagreement—the discussion about lethal autonomous weapons transforms. It is no longer a debate only about killing but also about dying. The core question shifts from an academic debate about morally preferable ways of taking an enemy's life to one about the principles for which a democracy is willing to lose its citizens in battle or—in the extreme—to lose a war. If warfare is going to change and the United States refuses to change with it, then it should not be surprised when the technological advantages that it has long enjoyed fade away, when the price of victory is higher, and when even victory itself is far from assured.[81]

The idea that people must at times be sacrificed for principles is acceptable in a democracy, even if it is unpalatable. The American military publishes extensive guidance on how its combatants must comply with the laws of war, though it knows that such compliance might sometimes reduce the effectiveness of certain battlefield operations overseas.[82] At home, while the reach of government surveillance is often vast, the United States is not a police state even when it comes to national security threats; the Foreign Intelligence Surveillance Act and its subsequent amendments, as well as other policy directives, are designed to provide mechanisms of democratic accountability and constraint, though the efficacy of those measures is a subject of debate.[83] There is ample room for criticism of the way in which this balance is struck on a wide variety of intelligence and military activities, but nearly everyone would agree with the idea that democracies should be willing to sacrifice some degree of security to preserve core principles.

The pressing question when it comes to lethal autonomous weapons is how this trade-off between principle and battlefield effectiveness will manifest, if at all. The capabilities of the technology itself are unclear, as many weapons systems are still in development stages. There are significant structural issues around bias and opacity with any AI system. More advanced future capabilities may also raise concerns related to specification gaming and a lack of understanding; for example, how will engineers train neural networks to make decisions in battle? As long as these weaknesses and questions persist, they will shape the geopolitical landscape and pose thorny challenges for warriors seeking to wield the new fire for war.

In the end, governments will likely resolve the questions around lethal autonomous weapons on the fly. Suchman believes this would be dangerous, regrettable, and avoidable. She and other critics of lethal autonomous weapons contend that a better understanding of the technology's limits would prompt caution and that no visible foreign threat justifies racing ahead. Given current trends, it seems likely that warriors in nations like the United States and China will continue to press onward over such objections from Cassandras, however. But moving forward does not necessarily imply disregarding ethical concerns in the development and use of lethal autonomous weapons. Democratic nations that invest in AI safety and security can perhaps mitigate the risks and improve understanding of the technology and its limits.

For Work, inventing new capabilities is wise, even if it means improvising on some key questions. If the United States waits until the uncertainty around lethal autonomous weapons is resolved, it will be too late. "The key thing is, where is our sense of urgency?" he said.[84] Work acknowledged the concerns of Suchman and others about what the technology cannot do. "But I'm more worried about us not pursuing this technology for competitive military advantage because it is clear that our competitors are doing it," he said again. "You don't want to be on the back end of a military technical revolution. All you have to do is ask the Iraqi Army how that felt."[85]

Work pointed to past eras of military transformation to suggest that, to some degree, the technology must be invented before its implications can be understood. Military strategists knew that aviation would transform naval warfare, he said, but couldn't predict how. Some thought the airplanes would spot targets for battleships to hit, while others thought airplanes would be an independent striking arm of the fleet. It wasn't until near the end of World War II that the US Navy formalized its carrier strike group concepts, which combine an aircraft carrier with cruisers, destroyers, and an air wing. "This was a transformation," Work observed. He didn't say it explicitly, but it was a transformation that the United States used to extraordinary advantage, leapfrogging other nations that had built battleship-centric navies for decades.

Michael Groen, the three-star Marine general who succeeded Jack Shanahan as the director of the Pentagon's Joint AI Center, also agreed. The military transformation that Groen wanted to avoid repeating was from World War I. At the time, commanders on both sides were surprised when their then-conventional infantry tactics proved no match for the raw power of machine guns, leading to tremendous bloodshed and stalemates on the battlefield. Groen noted that the new technology itself wasn't a surprise to the commanders, but they had refused to update their way of thinking accordingly. "It's up to us to act and act now," he said, "so that we don't miss this transformation, so . . . we don't have lancers riding into machine guns, or the Information Age equivalent of that."[86]

To adapt to a potential AI transformation of war, government strategists and ethicists will have to answer important questions. When and under what conditions is it permissible for an autonomous machine to take a human life? Will democracies be forced to cede a battlefield advantage to autocracies to preserve a moral high ground, or might they be forced to compromise on ethical principles to stay ahead? What will be the geopolitical implications—such as arms races and inadvertent escalation—when machines are making more decisions? And, perhaps the hardest question of them all, who will win?

We can only speculate about the answers. To the concerns of Cassandras like Suchman, Work and warriors everywhere offer the same rejoinder. "You're not certain where you will land," he said. "You know you will get there—you just have to go."[87] And so nations race ahead, perhaps transforming what is today unthinkable into what is tomorrow inevitable. When it comes to lethal autonomous weapons, it may before too long be December 8, 1941.

7

HACKING

At summer camps all across the United States, children play a game called capture the flag. In a field divided in half, players must steal a flag from their opponent's territory and bring it back to their own side while at the same time protecting their own flag from a similar theft. With its mix of offense and defense, the game tests players' abilities to plan and execute strategies that balance competing priorities.

At the DEF CON hacking conference—part of what is informally known as the Hacker Summer Camp, held every year in Las Vegas—a different version of capture the flag predominates. In air-conditioned hotel ballrooms, players must hunt for vulnerabilities in computer code, playing offense by exploiting those flaws in others' systems and defense by fixing flaws in their own. The game has evolved over decades to become one of the purest tests of hackers' mettle.

These software vulnerabilities matter because they enable modern hacking efforts. Hackers exploit vulnerabilities to cause software to do something it is not intended to do, such as reveal information, alter critical data, or stop working altogether. The US government spends billions of dollars each year trying to find and exploit vulnerabilities to enable offensive cyber operations and to find and fix vulnerabilities in code run by Americans.[1]

These offensive and defensive cyber operations are part of modern statecraft. The United States, China, Russia, and other nations try to break into one another's computer networks to advance their national interests every day. US government hackers have gathered intelligence, sabotaged rogue nuclear programs, and launched digital strikes against the Islamic State. Russian hackers have caused blackouts in Ukraine, interfered in elections, wreaked billions of dollars in damage around the world, and carried out extraordinarily broad and intrusive espionage campaigns. Chinese hackers have assembled giant databases with information on hundreds of millions of Americans and have stolen reams of economic and military secrets.[2] In each of these cases and many more, the pattern is apparent: hacking is a way for nations to gain an advantage over their rivals, an indelible aspect of geopolitical competition.

Hacking is about more than just exploiting vulnerabilities, however. Government hackers must intricately plan and execute their complex missions. They must learn about the targets they will strike, determine what software the targets are running, devise a method to deliver the exploit code that takes advantage of the software's vulnerability, and coordinate with other ongoing allied hackers to minimize inadvertent interference.

DARPA, the research arm of the Pentagon, wanted to know how AI could change cyber operations and assist American government hackers. Just as the warriors thought that lethal autonomous weapons would transform combat, many of them also believed autonomous digital systems would change cyber operations. A government that could automate the discovery of flaws in code would be better able to exploit its adversaries' networks. Similarly, a government that could automate key parts of planning and executing cyber operations would be able to carry out its campaigns more quickly and more effectively, staying one step ahead of the opposition. And a government that could use AI for cyber defense would keep its own secrets safe. In each task, DARPA wanted the leading government to be the United States.

In 2012 and 2013, DARPA began a multiyear effort to transform cyber operations and strengthen the United States' position online. As a result of the agency's programs and advances from other researchers, hacking has evolved to include more automation, some of it traditional and rule-based in its approach and some of it dependent on machine learning.

Cyber operations are a cat-and-mouse game between intruder and defender, and AI is changing both the tactics and the rules. What one side does to gain an advantage, the other side must learn to counteract; better and more automated offensive capabilities create a need for better and more automated defensive ones, and vice versa. Though humans will retain substantial roles in both offense and defense, the automation of cyber operations continues to improve, as warriors once more seek advantages for their nations.

But, as ever, the Cassandras raise an issue of vital concern: machine learning systems are themselves vulnerable to hacking. These systems are simply not as secure as many of the evangelists and warriors would sometimes like to believe. They contain all the kinds of regular software vulnerabilities that hackers typically seek and they also present new kinds of flaws. The very nature of neural networks renders much of modern AI brittle and vulnerable—a perfect target for sophisticated hackers in a new era of cyber conflict.

A DUEL WITH THE INFINITE

In an attempt to determine whether machines could find, fix, and exploit software vulnerabilities, in 2013 DARPA created what it called the Cyber Grand Challenge. The opposite of many traditional government contracts, which often feature exacting requirements and specifications, a Grand Challenge states an objective but does not instruct competitors on how to achieve it. It is up to the participants to devise their own strategy and means of execution. For the Cyber Grand Challenge, DARPA set the rules of the game, judged the results, and sought to determine which of the showcased technological innovations would help the Department of Defense succeed in its mission. DARPA offered a $2 million prize to the creators of the best automated method of finding, exploiting, and fixing software vulnerabilities, as determined by a final round of capture the flag to be held in Las Vegas at DEF CON in 2016.

Previous Grand Challenges in other fields had a good track record of kickstarting technological growth and shaping the future. The most notable challenge occurred in 2004, when DARPA offered a million-dollar prize to the team that created an autonomous vehicle that could navigate a 150-mile course. No one succeeded, but the challenge sparked a great deal of relevant investment and research. In a similar autonomous vehicle challenge the following year, nearly every team was better than the best team from the year before, and five teams finished DARPA's course.[3] Almost two decades later, some measure of autonomy is an increasingly common feature in cars.

To oversee the Cyber Grand Challenge, DARPA approached Mike Walker, an expert at capture the flag. His teams had reached the DEF CON competition's final round twelve times.[4] He had mastered the game's nuances: the intricate balance between offense and defense, the need to make the most of precious minutes and seconds, and the challenge of reacting quickly to the organizers' new wrinkles. Looking for software flaws in complex computer code, he said, was "a duel with the infinite."[5]

At first, Walker believed it was a duel only for humans; he thought it was impossible for a machine to play capture the flag. Impossible, however, was not a word that DARPA liked to hear, and after further prodding and cajoling, Walker agreed to give the machine competition a shot. He embraced the ambitious nature of the project, saying later that "a Grand Challenge is about starting technology revolutions."[6]

Walker and his DARPA colleagues created what in essence was a new computer operating system in which the automated hackers would hunt for flaws. The fresh digital terrain established a level field of play and ensured that no team had inside knowledge. As in human capture the flag hacking competitions, Walker and his team set up a mechanism through which a digital referee would give the entrants pieces of software to run and defend, some of which contained exploitable vulnerabilities. The automated competitors would have to determine what to do about these flaws without any human guidance whatsoever.

A complicated scoring system graded their performance along three dimensions. First, defense: the competitors had to find ways to fix the software that they were given, removing vulnerabilities and ensuring that other teams could not exploit them. Second, function: they had to make sure that their fixes did not break the software or unduly take it offline; just as in the real world, a software fix that removed a security flaw but made the code slow or useless would not count as a success.

Third, offense: competitors had to exploit other teams' software, finding and taking advantage of vulnerabilities in their code—the digital equivalent of stealing the flag.[7]

David Brumley, a professor at Carnegie Mellon and himself a top player of capture the flag, decided to enter the Cyber Grand Challenge. Brumley knew how important exploitable software vulnerabilities were to cyber operations and how much intense human effort was required to find them. Throughout his career, he sought ways to automate the task, even as some of the most well-known hackers in the world doubted that it would be possible to do so. The Cyber Grand Challenge was an opportunity to prove the skeptics wrong and show what automation could achieve. Perhaps in a sign of the change that he saw coming to the field, Brumley named his entry Mayhem.[8]

As a university professor, Brumley received a small grant from DARPA to fund his team. Even so, he didn't have a lot of money to spend on the project compared with other leading competitors, which included major defense contractors. Some of the people he wanted to hire could easily have commanded upward of $250,000 per year in the private sector, far beyond what Brumley could pay. Instead, he went to his top recruits with a reminder that the prize was $2 million and an offer that, if his team won, every member would receive their fair share. It was a clear incentive.[9]

Brumley's team began their effort by trying to use machine learning to discover software vulnerabilities. They found that, at least when they worked on the problem in 2014 and 2015, the technology wasn't up to the task. One of the major issues was that machine learning systems often improved by getting incrementally better at solving a particular problem, such as AlexNet training its neural network to recognize images more accurately over time. At the time, this kind of incremental process didn't work as well for finding software vulnerabilities, as it was often hard to tell if one was actually making progress toward the goal, a key requirement for iterative machine learning methods.[10]

For this reason, Brumley's team turned to more traditional methods of rule-based automation rather than machine learning for some parts of Mayhem. To find and fix software vulnerabilities, Mayhem used these heuristics to look for signs that a vulnerability might be present, perhaps in ways that could make the software crash. To fix the problem, it would rewrite sections of the software or tack on an addition to the end of the program, making sure to preserve as much functionality as possible. It was an impressive feat of automation. The ability to defend code while keeping it running as much as possible would earn Mayhem points for defense and functionality; Brumley called mastering this balance the most boring, but still vital, part of the Cyber Grand Challenge.[11]

To win, Mayhem had to play offense too. It used two largely independent approaches to develop code to exploit the vulnerabilities in other teams' software. First, it attempted to reverse engineer the logical structure of its opponents' code. It tried to figure out, for example, the conditions under which certain parts of the software would execute, and what flaws might exist in the logic. Second, a technique called fuzzing worked together with Mayhem's vulnerability-discovery tools to try to get the rival software to crash—often a sign of a flaw in the code. Once one of these two methods found a weak spot, another part of Mayhem generated code to take advantage of the uncovered flaws and give Mayhem the capacity to manipulate the targeted software.

These capabilities wouldn't amount to much if Mayhem did not make the right strategic choices. Just like a real-world hacker, Mayhem had to observe what its opponents were doing, understand their capabilities and decisions, and craft its own mix of offense and defense. To build this capacity, Brumley's team deployed a basic machine learning algorithm. The algorithm observed how the game unfolded, watched the actions of other automated players, and made assessments about which software fixes Mayhem should deploy. The adaptive algorithm gave Mayhem flexibility, enabling it to adjust its strategy as the game evolved and to stay alert to other teams' attempts at deception.

Machine learning also helped Mayhem deceive others. Brumley and his team realized that they could gain an advantage over their competitors by fooling them into thinking that Mayhem had found vulnerabilities in particular pieces of software when it had not. To mislead Mayhem's automated opponents, Brumley's team built a machine learning system that continually produced distractions. While the others wasted time and computational resources trying to sort through the machine-generated noise, Mayhem would try to find actual vulnerabilities, develop ways to exploit them for offensive purposes, and devise fixes that addressed the problems in the software it was defending.[12]

The final round took place in Las Vegas in August 2016. Spectators crowded into the ballroom to watch visual representations of the action, aided by an announcer who interpreted the game's events. Mayhem, in the form of a seven-foot-tall black box with glowing neon lights and rows of computer processors, got off to a strong start. Its approach for finding, exploiting, and fixing software vulnerabilities without unduly degrading functionality worked well. Even better, the more strategic elements of Mayhem excelled over its competitors. The system carefully chose how to balance offense and defense, outmaneuvering other teams that were less agile and deliberate in their approach. Its attempts at deception also succeeded; in one case, the announcer even informed the crowd that Mayhem had found a vulnerability when in fact it was a decoy.[13] As the morning continued, Mayhem's prospects were looking good.

Then Mayhem suddenly stopped working. Even years later, Brumley can't be certain what happened. A hard disk had failed, but Mayhem was built to overcome accidents like that. The more likely cause, he later concluded, was that "the most trivial" piece of Mayhem's code—the part that downloaded the software from DARPA's servers and submitted Mayhem's fixes back—had failed. It was not keeping up with the heat of competition. Mayhem was doing a lot of good work, but like a student turning in a paper after the deadline, it wasn't doing it fast enough to make the grade.

Brumley's team asked DARPA if they could reboot Mayhem. Walker said no. Fully autonomous meant fully autonomous, he reasoned, and no human intervention was allowed. Key members of Brumley's team apologized to one another, taking responsibility for the nonfunctioning embarrassment that Mayhem had seemingly become. Other team members started handing out the Mayhem promotional t-shirts they had brought, assuming that no one would want them after Mayhem's failure became more obvious and other competitors caught up. Brumley compared the helpless feeling to being "in a sports game when you're down 49 to 0 or something and you think you're just going to lose." He thought about all the years he had spent advocating for research into the autonomous exploitation of software vulnerabilities and all the big-name hackers who had dismissed his efforts. That work felt like it was "down the drain," he said later.[14]

Brumley and his team need not have worried so much. Three things worked in their favor. First, Mayhem wasn't actually down 49–0; it had built up a giant lead before it stopped working, and other teams were closing the gap only slowly. Second, because the dormant Mayhem never took its software offline for repair, it received some points for maintaining the code's functionality and availability. Third, very late in the competition, for reasons that are still unclear to Brumley, Mayhem's system for interacting with the DARPA server began working again.[15] When the dust settled, Mayhem had won.

Mayhem's victory attracted an enormous amount of attention. Walker said that he was "amazed at the speed with which the machines responded to the use of bugs in software they had never seen before and fielded patches in response."[16] *Wired* magazine proclaimed that "hackers don't have to be human anymore."[17] For as imperfect as Mayhem was, it was easy to think that the victory foreshadowed a new age of cybersecurity. Just as DARPA's Grand Challenge in 2004 was a harbinger of improved vehicular autonomy, the narrative went, this moment was a sign of improvements to come in cybersecurity autonomy.

Nearly every recounting of the Cyber Grand Challenge stops with Mayhem's victory over other machines. But these narratives omit the events of the following day, events that shed at least as much light on the potential capacity for autonomous systems to outperform humans when it comes to finding and exploiting software vulnerabilities.

The day after their triumph against other machines, an exhausted Brumley and his team prepared Mayhem to battle against the best hackers in the world in the human capture the flag competition. They hoped to achieve a parallel success to AlphaGo's victories over top humans. Unlike with AlphaGo, however, this showdown against humanity wasn't really a fair fight. The technical setups of the human and machine capture the flag competitions weren't the same, and Mayhem had been built for the latter. An unexpected software bug in the human capture the flag competition's virtual environment complicated things further. It was as though Mayhem had won a game of American football one day and then tried to compete at soccer the next. The disparity was obvious from the start. Walker said in his opening remarks, "I don't expect Mayhem to finish well. This competition is played by masters and this is their home turf. Any finish for the machine save last place would be shocking."[18]

Even so, the human hackers recognized that they were playing against a capable digital opponent. One of the participants was George Hotz—or geohot, as he was known in hacking circles. Hotz, whom Brumley called "the best hacker in the world," had a well-deserved reputation at DEF CON. At age seventeen, he had become the first person to hack the software constraints on his iPhone so that it could be used with any wireless carrier.[19] A few years later, he found vulnerabilities in the Sony PlayStation service that were so serious that the company took the entire system offline to correct them.[20] By the time of the 2016 competition, his ability to find and exploit software flaws was in no doubt.

As the human game of capture the flag unfolded, Hotz found something interesting in one of the pieces of code provided to competitors. It was a quirk that, as he told one of his teammates, looked like it would present a real challenge to exploit. As he got to work, all of a sudden Mayhem found the same vulnerability and quickly exploited it with a clever method of its own. The machine's solution was intricate and required several stages of corrupting the memory of the computer running the software. Though Mayhem finished last in the human competition, to some this triumph was akin to AlphaGo's move 37 against Lee Sedol—one more sign that machines could outperform humans, even in the complex world of software analysis.[21]

Walker said later that Mayhem's success at exploiting the tricky vulnerability—especially in an uneven competition where the playing field was tilted against it—was "probably the most interesting thing" that happened at the Cyber Grand Challenge even though it attracted little attention. That Mayhem had exploited the complex vulnerability was more notable than the machine's victory over other machines the day before, and certainly more important than its own eventual defeat in the human competition. Tools like Mayhem would only get better, enabling machines to aid cyber operations and freeing up humans to focus on other tasks. The success in devising the method of exploitation was, Brumley agreed later, "a sign of things to come."[22]

The question was what exactly would come next after Mayhem's victory. Brumley said that he thought the finalists at the Cyber Grand Challenge showed that fully autonomous hacking was possible. He and some of his teammates formed a company to use Mayhem's successors to find flaws in military code. To prove Mayhem's capabilities translated beyond the DARPA competition, Brumley demonstrated that the machine could find previously unknown flaws in the software of airplanes.[23]

But the answer to the question of what came next for AI in cyber operations did not depend just on Mayhem. Another secretive DARPA program, one that was less flashy than the Cyber Grand Challenge but at least as impactful, also imagined what autonomous hackers could do. It sought not just to enable hacking operations by finding vulnerabilities but to reimagine cyberwarfare itself.

PLAN X

The Pentagon knew that technology like Mayhem, however significant, was just one link in a broader chain of automation for cyber operations. Finding and exploiting software vulnerabilities was fundamental to modern hacking, but DARPA wanted AI to do still more, shaping not just how vulnerabilities were found but how American hackers used them against adversaries. AI could transform other parts of offensive endeavors, including the process of understanding target networks and planning missions to attack them.[24]

Offensive cyber operations depend first on understanding the ever-changing digital environment. Changes in software shape which vulnerabilities are present and how hackers can exploit them. The former National Security Agency director Michael Hayden once said, "Access bought with months if not years of effort can be lost with a casual upgrade of the targeted system, not even one designed to improve defenses, but merely an administrative upgrade from something 2.0 to something 3.0."[25] Other changes help determine which path across the internet hackers take to reach their targets; sometimes, it might be best to launch an operation via a satellite network, whereas in other circumstances it might make the most sense to sneak through the ground-based networks of another nation to reach the target. Sophisticated hackers must use their judgment and experience to make these kinds of decisions every time they go to work.

The complexity of the offensive process only increases when a nation carries out many overlapping operations at once. For years, American hacking teams often conducted reconnaissance of targets and wrote hacking plans on whiteboards in their operations centers.[26] Coordinating missions across hacking teams—often from different military and intelligence units—was even more cumbersome; American planners regularly had to determine if their operation would interfere with the activities of others.

In parallel to the Cyber Grand Challenge that sought to automate discovering software vulnerabilities, DARPA announced a separate effort in 2012 to improve the planning and execution of offensive cyber operations. The program was officially called Foundational Cyber Warfare, but everyone just called it Plan X. While the name sounded like it was drawn from science fiction, a subtle meaning was buried within. The X was a Roman numeral referring to Title 10, the legal authority under which the United States conducts military operations; it signaled that the military was trying to develop notable capabilities in an arena—offensive cyber operations—that had long been dominated by the intelligence community. Or, as DARPA put it, Plan X would try to get past what it called the "manual" way of managing cyber operations, which "fails to address a fundamental principle of cyberspace: that it operates at machine speed, not human speed." While speed was not the only thing that mattered in offensive operations—stealth and persistence were often also crucial for success—DARPA was excited enough about the prospect of hacking faster that it paid more than $110 million to get Plan X off the ground.[27]

Plan X was ambitious, aiming to link together key parts of hacking operations, from planning to reconnaissance to execution. In prototypes, glowing tabletop interfaces displayed maps with red dots representing adversary computers that could be hacked. Operators could tap on the dots and reveal the collected intelligence about the system. They could learn about the target's weaknesses and see which available exploits could take advantage of those flaws. Arati Prabhakar, who headed DARPA when it was developing Plan X, said the goal was to "allow cyber offense to move from the world we're in today—where it's a fine, handcrafted capability that requires exquisite authorities to do anything . . . to a future where cyber is a capability like other weapons."[28] In essence, by giving more autonomy to the hacking software, DARPA wanted military hackers to be able to operate more quickly and with less training, enabling greater operational scale.

Plan X didn't live up to its ambitions right away. While the futuristic demos offered a *Minority Report* vibe to the prosaic business of cyber operations, reality didn't always deliver on that vision. As Plan X's developers iterated through different designs and objectives, they retrenched to focus on narrower tasks, such as making sure different kinds of software for planning hacking operations could communicate. It was important work, but hardly a full-scale transformation of government hacking efforts.

The commander of the US Army Cyber Command, Lieutenant General Ed Cardon, thought the narrowed scope of Plan X was a missed opportunity. In a meeting in 2015, he implored the manager of Plan X, Frank Pound, to consider how Plan X could live up to its original mission of remaking cyber operations. He wanted Plan X to draw in all the data relevant to American cyber operations, offer insights on what kinds of efforts would succeed, and automate as much as possible of government hacking.[29]

The next year, Cardon became even more confident in his conviction when he attended the Cyber Grand Challenge at DEF CON in Las Vegas. He later said his reaction to Mayhem was simple: "I want one of those!"[30] He was "stunned" as he watched the machine find vulnerabilities in software and develop code to exploit those weaknesses. Immediately he thought about how cyber operations would change. "When I saw that, I'm like, 'Oh my gosh, if you put that together with Plan X, this is how you would conduct operations,'" Cardon said. Mayhem could someday work alongside other tools to more fully automate attacks in all of their phases, discovering vulnerabilities and writing the exploits that Plan X would help hackers deploy.[31]

At DARPA, Pound agreed that Plan X could do more. As he and Cardon worked with a group of staffers to tightly link Plan X to the Army's burgeoning cyber operations capability, the project entered a new phase. By 2018, the interface was more mundane—no more glowing tables—but also more robust. The new system showed which military hackers were focused on a target and how they could work together. As in the early prototypes, it offered a way for the hackers to learn about their targets and plan their operations. "Think of it as a full-spectrum cyber operations platform specifically designed for military operations," Pound said.[32]

The evolution only continued from there. In partnership with the Air Force Research Lab, the Pentagon's Strategic Capabilities Office—a secretive unit designed to turn powerful inventions, including those originating with DARPA, into war-fighting capabilities—took over Plan X in December 2018. The office renamed it Project IKE, a moniker meant to project a friendlier image for the advanced hacking effort and reportedly a play on Dwight Eisenhower's cheery "I Like Ike" campaign slogan. The details of what the new software could do were buried under new layers of classification, but there is no doubt that the technology continued to advance.[33]

The leadership of the Strategic Capabilities Office reportedly had one major priority for the continued development of Project IKE: to add more machine learning. Machine learning had come a long way since the Cyber Grand Challenge and was increasingly powering many other AI systems. The Pentagon wanted to ensure that Project IKE was making the most of the powerful technology, using it to assess how cyber operations would unfold, which computers would be worth hacking, and how hacks against those targets could be made more efficient. An updated version of Project IKE is reportedly rolled out to US Cyber Command—the military's top hackers—every three weeks.[34]

One of Project IKE's features is a way of forecasting what could happen in any given cyber operation. At least conceptually, this analysis is akin to the way in which AlphaGo and its successors forecasted how games of Go would develop, though hacking is far more complex. The system quantifies the chances of success for a given cyber operation as well as the chances that something might go wrong and cause collateral damage. Concerns about unintended consequences had bedeviled even sophisticated hacking efforts for years, especially for the attacks that relied heavily on early forms of automation to achieve a large scale. The most famous—or infamous—example was Stuxnet, the American and Israeli attack on Iran's nuclear program. Discovered in 2010, the code autonomously spread itself too aggressively and infected computers all over the world.[35] Project IKE's increased situational awareness and understanding of target networks aimed to give policymakers more confidence that such blowback wouldn't happen again.

The ability to quantify the chances of success and risk was a game-changer for a warrior like Cardon. He had led many cyber operations against the Islamic State and warned policymakers repeatedly of the uncertainty that surrounded modern hacking campaigns. Project IKE offered his successors the capacity to more fully understand what could go wrong, give senior leaders a grounded assessment of how likely success was, and win approval where appropriate to carry out the operation. "That was what was powerful," he said. "It categorized risk in a way that I could have a pretty good level of confidence."[36]

It's impossible to know how good Project IKE actually is at estimating risk in the complex world of cyber operations. The intersection at which IKE exists—of AI and cyber operations—is an area rife with hype, and the US military has said little in public about what the technology can do. More generally, getting something like Project IKE to work requires managing not just the machine learning system but also all of the more prosaic systems that feed into it, including those for collecting and storing data from disparate sources. Such mundane tasks are almost always hidden from view, but they are vital to implementing AI in large organizations.

Even if Project IKE performs well in tests, it is easy to imagine Cassandras warning about the ways in which machine learning systems might misjudge the complex digital environment of the real world. As chapter 4 showed, machine learning systems in many contexts have confidently produced risk assessments and quantifiable evaluations that later turned out to be wrong. Training data for significant cyber attacks is limited, making success more elusive still. Like any neural network, Project IKE's machine learning system faces challenges in explaining how it has reached its conclusions. The extent of these weaknesses is, like so much else about the program, highly classified—if it is known at all.

As the technology improves, policymakers will have to determine which of its capabilities merit their trust and which deserve skepticism. In general, the United States' posture in cyber operations has become more aggressive. Fewer bureaucratic restrictions constrain operations than in the past and a new strategy of "persistent engagement" aims to ramp up the tempo, ambition, and scale of American hacking. This aggressive approach included actively thwarting some Russian efforts to interfere in the 2018 and 2020 elections, in part by attacking the internet connections of Russian hackers.[37] How machine learning fits into the new posture remains to be seen, although Cardon and others believe that Project IKE is ready to contribute.

The most senior cyber official in the US government, NSA director and Cyber Command head Paul Nakasone, indicated before he took the job in 2018 that he understood the potential of AI in offensive cyber operations. Referring to the possibility of automating the attack cycle, he said, "It's really easy to say, 'I'm going to get on this network, achieve presence, have a persistent ability to go after whatever I do in the future.' But what if you had a machine that did that? That was able to rapidly find a vulnerability? Identify a vulnerability? Have an implant? Put down a persisten[t] presence? And do that at machine speed? You've taken away a ton of the work we do today."[38] Though Nakasone did not mention any specific government program, Project IKE seems to be one significant step toward bringing his vision to reality.

After two years of leading Cyber Command and the NSA, Nakasone continued to have AI on his mind. In a 2020 article coauthored with another top cybersecurity official, Michael Sulmeyer, he explicitly addressed the possibility that AI would empower future cyber attacks. "It is not hard to imagine an AI-powered worm that could disrupt not just personal computers but mobile devices, industrial machinery, and more," the pair wrote.[39] Whether the United States is pursuing such an automated attack capability or just worried about others doing so is as yet impossible to know.

There is no doubt that some authoritarian powers already know what automation in cyber operations can do. Russia and North Korea have both shown ample willingness to carry out highly automated cyber attacks, though not yet ones that use machine learning. In 2017, Russia launched an operation that deployed basic rule-based automation to spread itself rapidly around the world, wiping data from hundreds of thousands of computers and causing more than $10 billion in damage. A similar North

Korean attack in 2017 caused between $4 billion and $8 billion in damage.[40] Other sophisticated hacking groups have put machine learning to work in automating the spread of their code and evading cyber defenses.[41] Against such a backdrop, and alongside developments like Project IKE, it is easy to conclude that automation will generally make cyber attacks easier, faster, and more palatable for policymakers.

Meanwhile, still another capability was required to make the machine-versus-machine vision of the future come to life: automated cyber defense. DARPA—and many others—had plans for that too.

DEFENDING AT MACHINE SPEED

Wade Shen had never meant to work in cybersecurity. He was a pioneer in AI whose work focused on machine translation and natural language processing. In 2014, however, he came to DARPA's Information Innovation Office—which oversaw Plan X, the Cyber Grand Challenge, and other AI efforts—at just the right moment. It was, he later said, an "accident of history."[42]

Shen had heard enough of his colleagues talking about cyber operations to realize something counterintuitive: the targets of hacking efforts were often not just machines but humans. If a hacking effort could dupe a human into revealing a critical password or granting access to a particular system, there might be no need for advanced software exploits—and even if an operation did use such exploits, they were often loaded onto target computers by humans opening a malicious attachment or clicking a dangerous link. "People are very easy to co-opt," Shen said. "That problem has to do with how human beings react and trust each other. There's a fundamental way in which that kind of trust can be exploited that is significantly easier than exploiting machines."[43]

Shen realized that to protect computer networks, DARPA would have to find a way to protect humans as well as machines. Even if software engineers could find and fix many software vulnerabilities, humans remained a weak link, and hackers would continue to target them to gain illicit access to their computers and networks. But large organizations were composed of many individuals, and each one received many messages every day—all of which were potential vectors for hackers. Any additional effort to protect humans and defend against social engineering would have to be automated and scalable in order to keep up.

In 2017, Shen started the Active Social Engineering Defense program at DARPA. One of its goals was to determine which messages were legitimate and which were suspicious. "Being able to look at more data and examine that data in the context of various other communication histories and things like that gives you the ability to do better on the detection side," Shen said. "No one reads their email and looks at their headers to see whether the incoming message is valid," but machines could quickly verify such information to spot adversaries in action.[44]

Shen's other goal was more ambitious. He wanted to use data about users not just to protect those people by detecting operations against them but also to deceive an adversary's hackers (and, maybe someday, its automated hacking software). His idea had its roots in a well-known cybersecurity trick: the honeypot. To set up a honeypot, defenders create a lure—it may be a file, a computer, a network—that is fake but of interest to attackers. They then study the hackers' techniques as they operate against this pseudo-target.

For example, when the US government was investigating a series of espionage operations in the late 1990s, investigators created a file that appeared to contain state secrets. The hackers took the file back to their own network, at which point they were prompted to download a special piece of software to read it. They did, unaware that the software revealed their location—Russia—to the US government.[45] Other honeypots cause hackers to reveal information about how they plan to carry out an attack, shed more light on their identity, or simply waste their time by inducing them to plan operations against fake machines.

Shen thought that machine learning could create more and better honeypots. The very same data that enabled machine learning systems to spot anomalies and detect malicious messages could also be used to create a false image of normalcy for attackers. Shen wanted an adversary to work hard to trick a target into doing something like surrendering a password, only to eventually find that the target was a machine learning system in disguise. "There's an opportunity to design an engagement with an attacker to make them spend a lot more of their time on potential victims that look promising and make it much more costly to mount an attack on a well-defended organization," he said. "You can generate lots and lots of these fake honeypots within an organization. Or you can have a small number that are highly realistic. The prospects of some trade-off in this space working out are very high."[46]

Inevitably, though, hackers will still be able to deliver some of their malicious code to their target networks. The key question for network defenders is how quickly they can find this malicious presence and root it out. Historically, detection has been a tremendous challenge, in large part because of how much software users run. Microsoft estimates that its cybersecurity protections in Windows must make ninety billion decisions per day about whether programs are legitimate or malicious. It is a scale at which only automated tools can function.[47]

One analysis found that, in 2011, it took corporate, nonprofit, and government network defenders 416 days on average to detect intruders. Such a lengthy window gave attackers enormous freedom to develop their operations and copy secrets and—if they wanted to—attack. Thanks in part to more automated defenses and better coordination between human analysts and automated tools, many of which use machine learning, the time before detection has dropped dramatically. A follow-up study concluded that in 2020 the average time until detection was down to 24 days.[48]

Many companies claim to use machine learning in their cybersecurity products. The excitement around this application of machine learning is remarkable, with related venture capital investments increasing from $46 million in 2011 to over $1 billion in 2018.[49] Some of the hype is based in fact, as machine learning can already help detect a hacker's presence using at least three imperfect methods, each of which shows promise for future improvements.

The first approach is to gather data on what network intrusions look like and use that data to train machine learning systems. Just as AlexNet learned from the training data within ImageNet, machine learning systems can learn which characteristics indicate that a hacker has gained illicit access to a network, assuming that there are similarities between network intrusions. But while getting images to feed into ImageNet as training data was reasonably straightforward, though time-consuming, it is harder to get the right training data on attacks. Worse, data about computer networks is far more complex than the data contained in a single picture. DARPA undertook significant research in 1999 to remedy these issues and helped get this approach off the ground, but much more work remains to be done.[50]

Another method is known as anomaly detection. Rather than trying to build a picture of what hackers' behavior looks like, network defenders focus on identifying benign network activity. The defenders understand what is normal for their network, such as the times users log on, the kind of websites they visit, the kinds of files they transfer, and so forth. Once they have a baseline assessment of normal behavior, they train a machine learning system to look for activity that does not fit the mold and is thus more likely to be malicious. Automated systems can flag that activity for human review or can try to block it entirely. The challenge with anomaly detection, however, is that false positives can sometimes be plentiful, since computer networks are not static. An extreme example of this change is that the habits of users all over the world shifted during the COVID-19 pandemic.[51]

A third approach is to study the software that users run. Defenders have long compiled lists of identifying characteristics of code that is known to be malicious. These signatures can guide basic automated systems in detecting an intrusion. In response, hackers have become adept at writing code that automatically reshapes itself to evade rule-based tools.[52] Some machine learning techniques aim to find these shape-shifting pieces of malicious code by focusing on the underlying functionality or identifying commonalities that appear in different shapes of the same code.[53] The technology has shown strong promise, though the continued drumbeat of major hacking incidents proves that it is not a panacea.

And there is one additional wrinkle often raised by the Cassandras: What if machine learning systems themselves are not as secure as people assumed?

THE CAT-AND-MOUSE GAME EXPANDS

In 2012, several years before the Cyber Grand Challenge and before Ian Goodfellow devised GANs, he received an email from his doctoral advisor at the University of Montreal, Yoshua Bengio. Like Geoffrey Hinton,

Bengio had earned a reputation as one of the pioneers of machine learning.[54] In this email to Goodfellow, Bengio raised a limitation of neural networks that was of growing research interest and concern: they could be hacked.[55]

Bengio introduced Goodfellow to a paper in progress by Christian Szegedy, an employee at Google. Szegedy was among the first to discover that adversaries could tweak the input to a trained neural network in a way that no human could spot but that would cause the machine learning system to fail. Altering just a few pixels in an image of a school bus, for example, could cause a machine learning system to instead identify it as an ostrich, even though a human who compared the altered and unaltered pictures would see no discernible difference.

Goodfellow assisted Szegedy with some of the research for the paper. He helped coin the term "adversarial examples" to refer to inputs crafted to fool machine learning systems.[56] As the two researchers began to explore adversarial examples further, they were struck by what they found. Whereas some kinds of software vulnerabilities, such as the ones later exploited in the Cyber Grand Challenge, were straightforward to fix once they were discovered, adversarial examples seemed to arise from an intrinsic weakness of the neural networks themselves. Changing a neural network to defend against one adversarial example often made the network more vulnerable to another example.

Goodfellow and others knew that the structure of a neural network and the configuration of its parameters—determined by the training process described in chapter 1—shaped how data cascaded through it. In AlexNet or another image classifier, the input data arrived in the first layer of the network and then, depending on the strength of the connections between neurons, moved through the layers before arriving at the output layer. The output layer expressed the neural network's identification of the image; a system designed to tell tanks from jeeps would usually have two neurons in the output layer, one for each possibility. If the network classified a particular picture as a tank, the tank neuron would contain a high value and the jeep neuron a low one. Machine learning scientists and the training process for each neural network aimed to produce systems that reliably generated the right outputs for each input.

Goodfellow and others realized that hackers could turn this process on its head to craft adversarial examples that fooled neural networks. To dupe a machine learning system into thinking a picture of a tank was a picture of a jeep, a hacker could begin with a picture of a tank that the trained machine learning system correctly recognized. The hacker could then make slight perturbations to the image, changing just a pixel or two, and observe how the neural network responded. If the small change in input caused the neural network to be slightly less confident that the picture was of a tank—that is, if the output layer neurons had a slightly lower value for "tank" and a slightly higher value for "jeep" than they did before the change—the hackers could make another tiny change of a similar nature. They could do this again and again, subtly manipulating the image in ways nearly imperceptible to the human eye until it looked like something else, at least to the machine.

On the one hand, it is striking that such deception works at all. Adversarial examples are a reminder that machine learning systems do not have human methods of perception and cognition. A few pixels are too small for a human to see, but to a neural network they might make all the difference between a school bus and an ostrich or a tank and a jeep. The machine does not process the image the way humans do, nor does it recognize objects in the same way, and so minute differences in just the right places matter a great deal. Further, opaque machine learning systems have no meaningful way of explaining why those few pixels matter so much. Worst of all, they do not understand in any meaningful sense what school buses and ostriches are in the first place, nor why humans so rarely confuse the two.

On the other hand, deceptions that exploit quirks in perception and cognition are not new, and other optical and auditory illusions can easily fool humans. The neural network of the brain must make sense of the information it receives and adapt its processing to account for certain patterns. For example, the many frames of a film give the illusion of continuous motion when played in rapid sequence. Other illusions exploit the fact that the brain has adapted to a world that has three dimensions, even while the human retina only sees images in two dimensions. It is for this reason that two equally sized horizontal lines appear to be different lengths when laid over a pair of converging lines, such as in a drawing of

7.1 Adversarial examples that fool neural networks are conceptually similar to optical and auditory illusions that can fool humans. In this case, two equal-sized horizontal lines appear to be different lengths when laid over a pair of converging lines, which here resemble railroad tracks receding into the distance.

railroad tracks. This kind of illusion is deliberately crafted to expose the assumptions in human perception and cognition.[57] Adversarial examples that fool neural networks are at least conceptually similar to these real-world examples that trick humans, but they exploit different kinds of perception and data processing.

Drawing on these ideas, Goodfellow and others made major break-throughs in the study of adversarial examples.[58] What they found was cause for alarm. Many different kinds of machine learning systems were vulnerable, and an adversarial example that duped one system often duped similar systems, just as an optical illusion that fools one person often fools many others. This transferability meant that sometimes hackers did not need detailed knowledge of a target neural network in order to craft an example to dupe it.[59]

Researchers began regularly publishing different kinds of examples. They devised glasses that stumped facial recognition systems; ways of

7.2 Many different kinds of machine learning systems are vulnerable to adversarial examples. In this case, researchers devised glasses that duped facial recognition systems into thinking computer science students (pictured in the first row) were other people (pictured in the second row).

placing tape on stop signs to trick computer vision systems into thinking they were speed limit signs; and even a 3D-printed turtle that a well-known image recognition system perceived to be a rifle from every angle.[60] Other researchers found thousands of adversarial examples that occurred naturally, for reasons that are still unclear, such as a picture of a dragonfly that machine learning systems consistently identified as a manhole cover.[61] All told, researchers published more than 2,000 papers with adversarial examples between 2013 and 2021, many of them spurred by a contest Goodfellow helped organize in 2017.[62]

All of this research showed that adversarial examples posed major problems. "Most defenses against adversarial examples that have been proposed so far just do not work very well at all," Goodfellow said. Even "the ones that do work are not adaptive. This means it is like they are playing a game of whack-a-mole: they close some vulnerabilities, but leave others open."[63] The way in which the possibility of adversarial examples emerged from the structure of neural networks themselves meant that the problem seemed intractable, and yet much of the hype around machine learning technology continued undiminished. "This is really something that everyone should be thinking about," Goodfellow warned.[64] The US Air Force seems to agree; in late 2019, it began a $15 million per year program to study how adversarial examples will affect AI systems in combat.[65] If machine learning systems are in fact as hard to secure as Goodfellow fears, the Cassandras have a strong argument to make that their use in military contexts against tech-savvy adversaries creates serious risks of failure.

Adversarial examples are vivid, but they are not the only way for hackers to exploit neural networks. Another major risk is that hackers could interfere with the training of the neural network itself. For example, hackers could manipulate the training data given to the neural network using something called a data poisoning attack; the next chapter will discuss how data poisoning efforts carried out by individuals and automated programs can thwart efforts to combat disinformation. In addition, hackers can also use traditional techniques to manipulate databases containing training data. If hackers can control what the neural network learns, they can control how it acts.

Yet another kind of attack works in reverse. Instead of manipulating training data to change what the targeted neural network learns and does, hackers can manipulate the network to reveal what it knows. For example, hackers could coax a facial recognition system into revealing the face of a person in its training data.[66] This kind of attack is particularly valuable if the neural network was trained on secret information, such as efforts to use machine learning to analyze classified intelligence. Even if the machine learning system is trained in a secured environment, when its creators deploy it into the world, they might unintentionally reveal closely held secrets.[67]

In considering all the different weaknesses of machine learning systems, Goodfellow came to a stark conclusion: "attacking machine learning seems easier than defending it."[68] His observation, made fairly early in the age of modern machine learning, echoes the insight offered by a Department of Defense strategist named Irwin Lebow in 1980 at the dawn of the personal computing age. Lebow theorized that computer security would continue to be a fundamental problem because, although security would improve over time, the capability and complexity—and thus the vulnerability—of hardware and software would increase more quickly.[69] More than four decades later, it is obvious that Lebow was right. If Goodfellow, too, is proven correct, it will raise substantial concerns for the trustworthiness of machine learning in adversarial environments like war and cyber operations.

The field of machine learning cybersecurity is nascent. Bruce Schneier, a well-known cryptographer, compared its position today to that of cryptanalysis—mathematical code-breaking—in the 1990s: "Attacks come easy, and defensive techniques are regularly broken soon after they're made public."[70] A prominent machine learning security researcher at Google, Nicholas Carlini, painted an even bleaker picture. He compared machine learning cybersecurity to cryptanalysis in the 1920s; at the time, policymakers and technical experts knew that codes were important but grasped little more than that.[71] The urgent need to improve machine learning cybersecurity prompted DARPA to announce a project in 2019 that aspires to "develop a new generation of defenses to thwart attempts to deceive machine learning algorithms."[72]

In the near term, the obvious and often unfixable flaws in machine learning systems present major issues. In a commercial context, these weaknesses call into question the reliability and robustness of neural networks. They open up the possibility that hackers could trigger failure in systems that are core to a company's business, perhaps threatening to do so unless a ransom is paid. As firms increase their use of machine learning and as the technology continues to advance, the rewards for profit-motivated criminal hackers in exploiting machine learning systems will grow in turn.

Geopolitical competition raises the stakes still higher.[73] The adversaries in this arena are not just criminals seeking cash but well-resourced governments motivated by the ever-present struggle for any advantage over adversaries. These governments recruit some of the most technically adept minds in the world to discover software weaknesses they can exploit, and sometimes spend billions of dollars to build out the teams and organizations to do so. As the national security applications of neural networks continue to expand—from intelligence analysis to lethal autonomous weapons to automated hacking tools and beyond—government hackers will increasingly target these systems' unique weaknesses. Some of these future hacking efforts will be overt, perhaps taking vital systems offline at a critical moment. Others may be more insidious, undermining machine decision-making in ways that are hard to detect.

The question is what to do. Cassandras warn of these systems' limitations, while warriors will often see a competitive need to press ahead, securing one's own systems as much as possible and exploiting those of one's adversaries. For better or worse, the warriors are mostly in charge of this decision. And so the cat-and-mouse game of cyber operations will continue to grow in speed and force, subsuming new technology into the never-ending struggle between attacker and defender.

8

LYING

The letter to the editor that ran in India's *Patriot* newspaper on July 16, 1983, began with a bold claim: "AIDS, the deadly mysterious disease which has caused havoc in the U.S., is believed to be the result of the Pentagon's experiments to develop new and dangerous biological weapons."[1] The letter outlined the mysteries of the disease's spread and the horrors it inflicted upon its victims. The writer accurately quoted Pentagon and CIA statements from years before about the American biological weapons program, and referenced the CIA's aggressive medical experimentation with mind-altering substances in the top secret MKULTRA project. There was no name attached to the note, but it was signed by a "well-known American scientist and anthropologist" based in New York.

The letter was not what it seemed. It originated not with a scientist in New York but with the Soviet intelligence service KGB, which had helped fund and start *Patriot* two decades before.[2] In the words of one definitive history, the ploy was "a masterfully executed disinformation operation: comprising about 20 percent forgery and 80 percent fact, truth and lies woven together, it was an eloquent, well-researched piece that gently led the reader, through convincing detail, to his or her own conclusion."[3] According to KGB documents, "The goal of these measures is to generate, for us, a beneficial view in other countries that this disease is the result of out-of-control secret experiments by U.S. intelligence agencies and the

Pentagon involving new types of biological weapons."[4] The effort aimed to detract from American warnings about Soviet biological weapons and alarm US allies in the developing world.

The Soviet disinformation campaign met with some success. Though the letter did not immediately change the headlines in allied capitals, further Soviet efforts to push the bioweapon theory of AIDS dovetailed with the work of American conspiracy theorists and Soviet partner intelligence agencies in the years to come. In 1986, a British tabloid reported that AIDS was an American creation, citing the work of an academic who had been convinced by the KGB ruse, and newspapers in more than thirty countries eventually repeated the Soviet claim. The next year, as the story's credibility continued to grow through repetition, the Associated Press reported that a Soviet analysis had concluded that AIDS had escaped from an American weapons laboratory. That night, Dan Rather—then one of the most well-known journalists in America—repeated the Soviet conclusion for his audience of fifteen million Americans on the *CBS Evening News*. Decades later, the claim still made the rounds, including in a 2006 interview with Kanye West in *Rolling Stone*.[5]

The Soviet operation is a case study in effective disinformation: sowing a false or misleading claim in order to influence a target audience, leading consumers to what seems like their own conclusions, and working to amplify that claim while helping it gain credibility. Disinformation has been a common tactic for centuries. Ben Franklin forged and leaked a letter supposedly from the German King Frederick of Hesse to King George, intending to diminish the morale of Hessian troops aiding the British in 1777.[6] Centuries later, one of the most prominent defectors from a Soviet intelligence agency, Ladislav Bittman, estimated that the total number of disinformation operations during the Cold War was more than 10,000.[7]

This history offers context for modern disinformation. The 2020 election cycle in the United States saw disinformation emanating both from the highest levels of the US government and from foreign adversaries. Despite the modest efforts of tech companies to curtail disinformation, conspiracy theories and organizing by violent groups flourished on social media. COVID-19 and the aftermath of the 2020 election presented ample opportunities for the weaponization of information and will likely prove a harbinger of the world to come—a world in which disinformation originates not just from America's adversaries abroad but from officials and elected leaders at home.

The new fire burns online too. AI fundamentally alters how malicious actors generate and spread disinformation. Automated networks of fake accounts can push messages far more easily and with greater speed and scale than individuals creating their own. Machine learning–enabled microtargeting of content on the internet means that more persuasive messages are far more likely to reach their intended audience than when operatives had to rely on newspapers and academics. Instead of the slow forgeries of the Cold War, the generative adversarial networks (GANs) described in chapter 1 can rapidly make such messages far more vivid than the KGB's letter to the editor. Such tools seem likely to continue to proliferate widely despite growing efforts to detect and control them.

Disinformation is a major geopolitical concern. NSA Director Paul Nakasone called it the most disruptive threat faced by the US intelligence community. "We've seen it now in our democratic processes and I think we're going to see it in our diplomatic processes, we're going to see it in warfare. We're going to see it in sowing civil distrust and distrust in different countries," he said.[8] The boundaries between the producers and consumers of information, and between domestic and foreign audiences, are blurred. The number of actors has multiplied, at home and abroad. Automation has accelerated the pace and potency of operations. And all of this is happening in the post-truth age, dominated by high levels of inequality and nativism, offering a more receptive audience than ever before.

THE WEDGE

When Rand Waltzman was a PhD student in computer science at the University of Maryland in the 1980s, his doctoral advisor told him to call an expert at a US government agency. In his research, Waltzman was interested in using computers to understand text, like many others at the time, but with a difference: he wanted to use computers to generate words, evaluate arguments, and then rip them apart. He was pushing into the seemingly uncharted territory of automated truth and lies. His professor had indicated that this expert, whose identity Waltzman did not know, was interested in similar subjects. But the call did not go as expected.

"Don't do it," the voice on the other line instructed when he heard of Waltzman's proposed research project.

"What do you mean?" Waltzman asked. "Do you think it's a bad idea?"

The voice responded bluntly, "Just don't do it."

Waltzman pressed. "What's the issue?"

The voice replied: "We don't do that sort of thing. And if we did, you'd never find the people who did it. And if you found them, they'd never talk to you."[9]

During the call, Waltzman reiterated his case for why the research was important, to no avail. The denials and obfuscations by the expert on the phone only deepened his interest. It was hardly the last time he would be told that his research was pushing on a closed door. He knew that others with less benign intentions would try to pry that door open, and he was determined to get there first.

Waltzman had detected both the potency and the continuing evolution of automated disinformation earlier than most. Whereas operatives had once carried out labor-intensive efforts in a slow and painstaking fashion, by the early 2000s, the internet was enabling higher-tempo efforts that were disjointed but effective. The automation of such disinformation campaigns would be next. No matter the medium or the origin of the message, however, the goal of disinformation remained the same: to deploy a potent mix of facts, forgeries, and fictions that exacerbated preexisting tensions, fissures, and contradictions in a society. These campaigns were not always lies cut from whole cloth but often amplifications of damaging messages or beliefs that were already in circulation. The best disinformation widened the cracks that already existed; it was a wedge.

In 2010, after two decades in academia and the private sector, Waltzman took a job at DARPA. He set up a program to study how information flows over social media, eventually receiving more than $50 million in funding. Building on his early work, he proposed to develop automated support tools that would include the capacity to detect and track ideas online, a classification system for recognizing disinformation campaigns, a method for identifying participants and their intent, and proposals to counter enemy messaging. The program resulted in more than 200 research papers from a range of authors exploring the possibilities of what Waltzman called "weapons of mass disruption."[10]

Of major concern were bots, automated accounts that mimic human operators. One study of Twitter links to notable websites suggests that more than two-thirds are shared by bots.[11] Some bots are harmless, if annoying. Spambots, for example, push commercial content to users and feature new products. Chatbots are designed to mimic real-life conversation. Other bots can serve as schedulers or handle customer service requests. Still others can alert the public to disasters as soon as they are recorded, such as the Twitter Earthquake Robot (@earthquakeBot) that warns about seismic activity.[12] Then there are the bots that act as social influencers, some of which peddle false or misleading content to advance a particular agenda. DARPA-funded researchers described them as "realistic, automated identities that illicitly shape discussion on sites like Twitter and Facebook."[13] Waltzman wanted to automate detection of these malicious bots.

His experiment began in February 2015, more than a year before the Russian election interference effort made disinformation a household term. Waltzman's program at DARPA funded a four-week challenge to develop techniques for identifying—and ultimately eliminating—influence bots. His team created a simulation of Twitter, complete with an artificial interface, more than 7,000 redacted user profiles, and 4 million tweets taken from actual people and bots who sought to change views during a 2014 debate about vaccinations.

The six teams that entered the challenge had to use an automated technique to find the bots supporting the pro-vaccination influence campaign. They had to differentiate them from bots spreading spam, bots aiming to make money, bots that were neutral or opposed to the campaign, and a host of other distractions.[14] Each team could register their program's guesses about which bots operated as social influencers and which were designed to perform other tasks. The teams would earn one point for each correct guess and lose a quarter of a point for each incorrect one. They could pocket additional points for the speed at which they correctly identified the bots. The winner would be the first team to find all the bots supporting the campaign.

The teams used a variety of techniques to accomplish this task. The winning team, from a Bethesda, Maryland–based company called Senti-Metrix, devised a clever strategy that blended both human and machine learning capabilities. The team first used unsupervised learning methods to cluster the accounts in the DARPA challenge into groups. With these groups identified, it used tweets about the 2014 Indian election—around which there was also a large amount of bot activity—as training data to make a guess about whether each of the identified groups contained bots. SentiMetrix deployed supervised learning systems to use the patterns of bot behavior it found to identify the bots in the challenge. All of this information and analysis was displayed in a dashboard for humans to examine and oversee.[15]

One of the key findings from this competition was that—contrary to the hopes of some evangelists—technology was not enough; human judgment was central to the goal of identifying and eliminating disinformation bots. "All teams used human judgment to augment automated bot identification processes," the researchers who supervised the competition observed.[16] As with the human-machine teaming discussed in chapter 4, user-friendly interfaces and timely feedback were critical to effective bot detection. Even as machine learning improves, the researchers supervising the competition concluded, it will not be sufficient to combat disinformation on its own. "Because new bots will be generated by adversaries who use different (and increasingly sophisticated) bot generation methods, we believe that machine learning by itself would be inadequate," they wrote.[17]

Even though SentiMetrix met with success in Waltzman's test, the Twitter bot challenge provided a sobering warning of a world of automated disinformation that seemed to grow more problematic by the day. Waltzman left DARPA in 2015 to take a job at Carnegie Mellon University. From his new perch, he watched disinformation campaigns achieve new heights in the 2016 presidential election. Waltzman worried that the failure to combat Russian interference that year—including via the automated defense tools rolled out by internet companies—was a sign of things to come. In particular, he feared that democracies would continue to struggle against disinformation. "We are waging this fight with two hands bound behind our backs, our feet tied, and our bodies chained to the wall," he said.[18]

The problem is both social and technical. Democracies provide an open and largely unencumbered environment through which even foreign actors can spread ideas. Cold War operatives had exploited that structure for decades; bots and fake accounts offer even easier access and broader reach.[19] But the role of AI is deeper still. The technology not only helps enable malicious messages but also shapes the digital landscape over which these messages travel.

THE TERRAIN

In 1996, not long after Amazon's founding, Jeff Bezos hired an editorial team. The team's first job was to write reviews of books, recommending them to customers in the clever and literary fashion of independent bookstores. As Amazon grew to selling more than just books, the team's remit expanded, and its members began to craft lighthearted recommendations for other products too. For example, in 1999, the Amazon home page declared of a children's lion-shaped backpack: "We ain't lion: this adorable Goliath Backpack Pal is a grrreat way to scare away those first-day-of-school jitters."[20] These recommendations were a whimsical and human touch in an era in which buying things online still often felt impersonal and intimidating.

Though Bezos spotted the need for welcoming warmth on his site, he also trusted data. He started a separate team of data scientists and engineers and tasked this team with recommending products to customers based on the other products they had bought or considered buying. This group eschewed quirky puns and playful descriptions in favor of automatically generated suggestions presented in a staid and standardized layout. The team saw its mission as nothing less than proving the supremacy of automation over humans at yet one more task: knowing what customers wanted. A sign in the team's office proclaimed "People Forget That John Henry Died in the End"—a reference to a folk legend in which a man tasked with driving steel into rock to build railroads outperforms a steam-powered machine, only to perish from the exertion.

The editorial team succumbed too. After reviewing the data, Bezos and his executives found that the personalized algorithmic recommendations drove more sales. Before long, the editors and writers transferred within the company, quit, or were laid off. In their stead came a recommendation system called Amabot. Not everyone liked the change; an anonymous Amazon employee took out a short advertisement in a Seattle newspaper, writing to Amabot, "If only you had a heart to absorb our hatred."[21] Twenty years later, however, that anonymous employee is almost surely gone from Amazon, while algorithmically sourced personalized recommendations remain.

The Amabot story reflects the evolution of the internet in miniature. What was once a place mostly governed by quirky humanness has become—with some exceptions—a more standardized ecosystem governed by machine learning. AI helps determine the ads users see, the tweets they come across, the search results they get, the videos they watch, the social media groups they join, and the posts from friends they read. As *Wired* magazine wrote in 2019, "The internet is an ocean of algorithms trying to tell you what to do."[22]

For disinformation operations, the implications of this change are profound: whereas malicious messages once spread via newspapers or broadcasts, in the modern age they travel across a terrain that is shaped both by the actions of humans and by the decisions of algorithms. The most effective modern disinformation campaigns are often the ones that go viral, manipulating both humanity and machines into exponentially fanning the flames of their message.

Facebook offers a prime example of the shifting terrain for disinformation operations. The company's head of applied machine learning, Joaquin Quiñonero Candela, told a group of engineers that "Facebook today cannot exist without AI. Every time you use Facebook or Instagram or Messenger, you may not realize it, but your experiences are being powered by AI."[23] Facebook's algorithms shape what content appears on users' feeds and thus what is most likely to go viral. With several billion users, it is hard for Facebook to determine the best mix of content for each user. As a result, the company constantly tweaks its algorithms to take a variety of factors into account, including the popularity of posts, the reactions of users, the credibility of news outlets, and more. It is a never-ending project with profound consequences for the flow of information online.

These choices matter for disinformation operations because news feeds and Facebook groups are where much of the action unfolds. They offer fertile terrain for divisive content. An internal Facebook study in 2018 concluded that "our algorithms exploit the human brain's attraction to divisiveness." Without substantial mitigations, these algorithms would feed Facebook users "more and more divisive content in an effort to gain user attention & increase time on the platform."[24] Warriors in autocracies have found that the automation-centric online environment works to their advantage.

The 2016 election offers an obvious case study. During the election cycle, Russian operatives set up groups on Facebook designed to spread their divisive messages as far as possible. These groups included Secured Borders, Blacktivist, United Muslims of America, Army of Jesus, Heart of Texas, and many others. By Election Day of 2016, many of the groups claimed hundreds of thousands of members; some were Russian operatives with fake accounts, while many others were Americans.[25] How any given American discovered these groups is impossible to know, but amplification as a result of Facebook's algorithms almost certainly played a key role.

Drawing in part on this history and on other examples, the internal Facebook study in 2018 specifically warned the company about the ongoing dangers of algorithms recommending groups to users. It found that 64 percent of all the people who joined extremist groups were steered to those groups by Facebook's algorithms.[26] An engineer who worked on the system identified it as one of his biggest concerns about Facebook. "They try to give you a mix of things you're part of and not part of, but the aspect of the algorithm that recommends groups that are similar to the one that you're part of is really dangerous," the engineer said. "That's where the bubble generation begins. A user enters one group, and Facebook essentially pigeonholes them into a lifestyle that they can never really get out of."[27] Partly in response to concerns like these, Facebook quietly turned off its recommendation algorithm for political groups in the run-up to the 2020 election, though apparently some political group recommendations nonetheless remained.[28]

Advertising is key to Facebook's operations—and the disinformation the platform is capable of spreading. Using a variety of methods, including tracking the sites users visit when they are not on Facebook, the company gathers an extraordinary amount of data about individuals all over the world. It then uses machine learning to segment users into a nearly limitless set of custom groupings, such as those with common interests in sports, organizations, or products. Advertisers can upload their own list of targets or customers and Facebook's machine learning systems will find others like them on its platform.[29] During the 2016 election interference operation, Russian operatives purchased more than 3,500 Facebook ads, many of them illegally; it remains to be seen how, if at all, disinformation operations of the future will use this capacity for ads targeted with machine learning.[30]

All told, disinformation created by Russian operatives persistently went viral throughout the 2016 election cycle, a sign of the struggle that seems certain to come with future campaigns. An after-action review by Facebook found that more than 126 million Americans had seen content that originated in the Russian campaign; a later study found that, in general, Facebook spread fake news faster than any other platform.[31] It is impossible to know how much of this content spread because Russian operatives sought to make things go viral, because American citizens shared messages with which they agreed, or because machine learning systems tried to show users the content they wanted to see. Disentangling the roles of humans and machines in assessing internet virality is a challenge for both companies and those who seek to spread disinformation.

Facebook has tried to fight back. Since 2016, the company has on more than one hundred occasions blocked users and closed down groups engaging in what the company calls "coordinated inauthentic behavior," with each takedown often removing many users and groups. The motivation of these groups varied, but some were composed of foreign government operatives aiming to sow disinformation, spread it widely, and advance their national interests. To combat this threat, Facebook also made changes to its algorithm to try to reduce the spread of content deliberately and deceptively crafted to go viral, though it rolled some of these changes back after the 2020 election.[32]

Major concerns remain about the platform's ability to fight disinformation. An audit of Facebook commissioned by the company and released
in 2020 raised substantial alarm about a variety of sources of misleading
information, including overseas actors. The consequences of amplifying
such information, from hate speech to other divisive material, threatened
the civil rights of the people subjected to this content and raised concerns
for the auditors, who wrote that the company had not done enough to
counteract the way in which "algorithms used by Facebook inadvertently
fuel extreme and polarizing content."[33] Some of this content is spread by
foreign actors operating through fake accounts; other content is spread
by domestic actors operating in their own name, often with a shrewd
understanding of what Facebook's algorithms will amplify.[34]

Facebook believes that the best way to moderate the content on its
platform is through more AI, not less. This approach fits within the evangelist hopes for the technology and all the problems that it can solve. The
company has long made clear its vision that machine learning will help
enforce its terms of service, which forbid not just coordinated disinformation campaigns but also hate speech, graphic images, the glorification
of violence, and other harmful material. Facebook relies on more than
15,000 human moderators to try to police its platform, but it has steadily
increased the role of machine learning in content moderation. In late
2020, for example, it rolled out a series of algorithms designed to better
determine which flagged content was breaking the rules, the severity of
the violations, and the likelihood that the content would go viral.[35]

The problem is not limited to Facebook. Google's YouTube has more
than two billion users across the world, greater than the number of
households that own televisions. Collectively, YouTube users watch more
than a billion hours of video every single day.[36] A user will often find the
next video to watch based on the company's recommendation algorithm,
which makes hundreds of millions of recommendations daily. It is this
machine learning system that guides users on their journey through the
video-sharing site; YouTube calls it "one of the largest scale and most
sophisticated industrial recommendation systems in existence."[37] It is
also a natural target for disinformation.

The tactics designed to make certain videos go viral can guide users
to dangerous places. For example, in 2017 and 2018, twenty-six Russian

state-funded media channels on YouTube drew more than nine billion total views, including many originating in YouTube's recommendation system. Among the stories posted on the Russian channels were claims that US politicians were covering up organ harvesting rings and that Scandinavian countries were on the brink of economic collapse.[38] Such success for Russian disinformation on YouTube continued a long-running trend; *RT*, a Kremlin-funded media outlet, became the first news channel to rack up more than one billion total views on the online platform back in 2013.[39]

Disinformation campaigns often tried to shape the public perception of breaking news events by manipulating YouTube's recommendation algorithm. For example, after the release in 2019 of the Mueller report, which provided the most authoritative treatment at the time of Russian interference in the 2016 US presidential election, Russian operatives sprang into action. They quickly placed a counter-narrative on *RT America*, Russia's state-funded propaganda outlet in the United States. In their video, American journalists criticized the press coverage of Mueller's conclusions, calling other reporters "Russiagate conspiracy theorists" and "gossiping courtiers to the elite."[40]

Even more remarkable than the disinformation about the Mueller report was the way in which YouTube's recommendation system drove traffic to it. A former YouTube engineer, Guillaume Chaslot, sampled recommendations from 1,000 YouTube channels and found that YouTube's system recommended the *RT* video more than 400,000 times; since the analysis drew from only a sample of channels, the actual number of recommendations for the video was likely "far higher." The *RT* video received automatic recommendations from more channels than any other video in the study, including those from major American news outlets. The analysis suggested that it was a sign that Russian operatives had become adept at fooling Google's algorithms, disproportionately guiding users—no matter their channel of origin on YouTube—to disinformation. "Every time they make something more pro-Russian, they get more and more views," Chaslot concluded.[41] In a different interview, he said, "The recommendation algorithm is not optimizing for what is truthful, or balanced, or healthy for democracy."[42]

For YouTube, like Facebook, the chosen remedy to such problems is more machine learning. The company has battled for years to upgrade its recommendation system's defenses against both foreign and domestic manipulations and to avoid driving users toward extremist, divisive, and false content. For example, in an effort to reduce the number of times the system recommended sensationalistic videos, the company altered the algorithm in 2016 to no longer prioritize maximizing the amount of time users spent on the site.[43] Instead, the algorithm prioritized user responses to videos. That enabled coordinated actions, sometimes by humans and sometimes by bots, to manipulate the algorithm's training data through collective feedback on videos.

In 2019, Sundar Pichai acknowledged that the cat-and-mouse game was ongoing, not just for YouTube but for all of Google. Referring to the threat of bots manipulating algorithms, he said, "This is something we actually face across the set of products today, be it our ad systems, be it our search products . . . be it YouTube and so on."[44] Despite employing many of the most savvy machine learning scientists on Earth, it was not clear that the companies had an advantage—and that was setting aside the fact that Ian Goodfellow's invention of GANs unintentionally made the challenge of taming the fires of falsehood even harder and the potential dangers of disinformation even graver.

THE LEVER

When France fell to the Nazis in 1940, Hitler insisted on applying a chilling symmetry: he ordered the French to surrender to Germany in the same railway car in which the Germans had conceded defeat to the Allies in World War I. As Hitler emerged from the compartment, he took what appeared to be a deliberate step downward, as if to emphasize that he had squashed the French under his boot. Cameras at the scene captured the moment. The Scottish filmmaker John Grierson later saw the footage and, working with another filmmaker, manipulated the tape to make it appear as if the Nazi were dancing a jig, insane with glee about his wartime victory. The edited video featured prominently in Allied propaganda, which used it to suggest that Hitler was an uncontrollable madman. Grierson was a celebrated filmmaker who coined the term "documentary" and helped to develop the genre; it is ironic that one of his most notable films took such license with the truth.[45]

Whether shooting documentaries or wartime propaganda, Grierson's work centered on an important theme: the moving image is powerful. The pen is mighty and pictures are worth a thousand words, but there is something unmatched about watching the action unfold. Viewers feel like they are in the scene, reliving what is happening and connecting to the moment.

This immersive, lifelike quality is what imbues deepfakes with such great potential power. Deepfake videos are often created by the GANs described in chapter 1; as discussed, these are machine learning systems that feature two neural networks working in competition with one another. One of the neural networks specializes in producing outputs that mimic, but do not replicate, a certain kind of real-world data, such as video of a particular person speaking, while the other network specializes in distinguishing products of the first network from the real-world data. Over time, each improves at its role and, when the system is trained, the first network can produce convincing fakes, including videos that seem to be real. A GAN could do easily and more convincingly what Grierson had to do painstakingly and manually.

Right now, deepfakes are most commonly used for purposes unrelated to geopolitics, such as swapping celebrity faces in videos or creating non-consensual pornography. But deepfakes have already been used for disinformation and will likely continue to appeal to disinformation operators of the future; the FBI warned in 2021 that China and Russia will "almost certainly" put them to malicious use.[46] In a deepfake, a politician might appear to say something racist or mentally unsound, or a military commander might order troops to surrender. Enemy soldiers might appear to commit human rights abuses or war crimes. A dissident who threatens a government's hold on power may appear to do something hypocritical. While lying and deception have always been part of the struggle for geopolitical advantage, deepfakes seem poised to take them to the next level.

The FBI warning in 2021 was not the first one. In their worldwide threat assessment, US intelligence agencies concluded in 2019 that "adversaries and strategic competitors probably will attempt to use deep fakes or similar machine-learning technologies to create convincing—but false—image, audio, and video files to augment influence campaigns directed against the United States and our allies and partners."[47] An early example of such an effort boosted the Chinese telecommunications giant Huawei, which many analysts believe has strong ties to the Chinese government. As European nations debated whether the company was eligible to win billions of dollars of contracts for their 5G telecommunications networks, Twitter accounts appeared online arguing on the firm's behalf. These accounts, which promoted an article written by a Belgian lawyer working with the company, appeared to be from ordinary people. Each account was complete with a fake profile picture—generated by a GAN—as well as a realistic-seeming Twitter biography. Huawei executives in Europe amplified the messages as a sign of the company's supposed grassroots support. The effort was part of a broader campaign that included opinion pieces in online publications, some of which also used GAN-generated images for author photos.[48]

This campaign is notable as a potential sign of things to come, but it is relatively minor compared with some of the other theorized harms of widespread deepfakes. One oft-cited concern is that deepfakes will cause the public to stop believing what it sees. Some call this effect the "liar's dividend," postulating that it most helps those who trade in falsehoods or are quick to label opponents as "fake news."[49] The denial of truth has long been a feature of autocracies. In her seminal study of totalitarianism, Hannah Arendt wrote, "Before mass leaders seize the power to fit reality to their lies, their propaganda is marked by its extreme contempt for facts as such, for in their opinion fact depends entirely on the power of man who can fabricate it."[50] Stalin and other autocrats rewrote history to preserve their hold on power, forcing citizens to accept what many knew was not true. In *1984*, his fictional depiction of a totalitarian state, George Orwell wrote, "The party told you to reject the evidence of your eyes and ears. It was their final, most essential command."[51] It is easy to imagine deepfakes enhancing this power, especially at a time of already-decreasing trust of traditionally credible sources of knowledge, such as newspapers of record and government experts.

Deepfakes exacerbate a long-standing social problem. Lies, even when poorly told, have always spread throughout societies. While deepfakes seem more real than what came before, some argue they will not necessarily be more effective. So-called "cheap fakes" or "shallow fakes"—videos that lack the sophistication of those produced by high-end GANs—already often succeed at going viral, suggesting that it is not the editing technology itself that matters. For example, in 2019 and 2020, doctored videos of House Speaker Nancy Pelosi apparently slurring her words at a news conference circulated widely on the internet, especially in conservative Facebook groups and on conservative media sites, where they received millions of views.[52] Even simpler was a story widely circulated before the 2016 election that falsely claimed that Pope Francis had endorsed Donald Trump.[53] While it is impossible to know how many people who saw the story took it into account when casting their ballot, it had almost one million engagements, such as likes and shares, on Facebook.

Nor are such rudimentary manipulations limited to domestic politics. Waltzman offered an example of a case that shaped his understanding of disinformation in combat. In March 2006, American Special Forces targeted an Iraqi militia, Jaish al-Mahdi. The operation killed sixteen members of this armed group, destroyed their weapons, and rescued a hostage. All evidence indicates that it was a successful mission carried out in an ethical way. "In the time it took for the soldiers to get back to their base— less than one hour," Waltzman later testified to Congress,

Jaish al-Mahdi soldiers had returned to the scene and rearranged the bodies of their fallen comrades to make it look as if they had been murdered while in the middle of prayer. They then put out pictures and press releases in Arabic and English showing the alleged atrocity. The U.S. unit had filmed its entire action and could prove this is not what happened. And yet it took almost three days before the U.S. military attempted to tell its side of the story in the media.[54]

This example foreshadowed the disruption to come, Waltzman thought, especially as the jihadist group members were able to observe the online reaction to their ploy. "This incident was one of the first clear demonstrations of how adversaries can now openly monitor American audience reactions to their messaging, in real time, from thousands of miles away and fine tune their actions accordingly," Waltzman said.[55] In this case, the problem was not technical but social: the wedge had found a crack. It did not matter if the fakes were shallow and detectable. Even when told clearly and fairly quickly, the truth was on the defensive.

While cheap fakes seem bound to sow havoc, the dangers of deepfakes—and the hunt for technical solutions to detect them—still keep sharp minds awake at night. Matt Turek is one of them. As the head of the DARPA program that is responsible for combating deepfakes—called Media Forensics, or MediFor—Turek felt a sense of urgency. "When MediFor started in 2016," he recalled, "the digital imagery playing field favored the manipulator."[56] By the time Turek started at DARPA in 2018, it was clear that his mission was to even things out, finding ways to spot videos generated by GANs and reduce the liar's dividend. Whatever method he devised had to be automated and fast, capable of working before the videos went viral. It was a daunting task.

Turek is fond of a saying from Archimedes: "Give me a lever and a place to stand, and I shall move the world." This quote reflected the kinds of problems he chose to work on over the course of his career in both the private sector and at DARPA, as well as the specific task of mitigating deepfakes. What mattered was getting into the right position, then developing and applying the right tools for the job. Internet platforms like Facebook and Google offered the optimal ground on which to stand when combating deepfakes; these platforms would be able to block the fake videos once they were detected. Turek just had to develop the lever, the means of detection.[57]

His team at DARPA funded researchers at the University of California, Berkeley, who in turn partnered with Adobe, the creators of widely used photo and video editing software. The researchers collected and algorithmically altered images from the internet to create a database of manipulated pictures. They also hired an artist to modify certain images further, thereby introducing an element of human creativity into the mix. By using image pairs, one real and one edited, as training data, the researchers taught a neural network to spot the differences. The results of one test were striking: human eyes achieved a 53 percent success rate in detecting fakes, while the algorithms sometimes managed as high as 99 percent.[58]

In the deepfake detection business, however, all progress is fleeting, and the tension between manipulator and detector continues. For this reason, Turek created DARPA's Semantic Forensics effort, or SemaFor, as a follow-on to MediFor, announcing contracts totaling more than $35 million in 2019 and 2020.[59] SemaFor aims to build tools to analyze deep-fakes and spot logical inconsistencies in videos. These flaws, like unnatu-ral blinking, mismatched earrings, and uneven backgrounds, are small and subtle, but they reveal that a video originated with a machine learn-ing system that did not fully understand the reality it sought to emulate. For example, SemaFor can search for inconsistencies between where the metadata claims the video was shot and what the content of that video actually shows. By creating algorithms that can detect these aberrations, SemaFor seeks to give human analysts a set of tools to defend against deepfakes and sometimes even to tie the fakes back to their creators. In essence, SemaFor hopes to turn the failures of machine learning dis-cussed in chapter 4 against those who wield the technology for purposes of disinformation.[60]

MediFor and SemaFor are not alone in trying to solve the problems of deepfake detection. A consortium of technology corporations led by Face-book and Microsoft teamed up with a range of civil society organizations and academics to launch the DeepFake Detection Challenge in 2019.[61] In 2020, the group released 100,000 deepfake videos as training data for machine learning systems trying to detect such videos. The videos offered a broad sample of the kinds of deepfakes posted online, including clips from more than 3,000 actors that featured a range of techniques for swap-ping faces or otherwise manipulating film. It was akin to ImageNet for deepfake detection.

More than 35,000 different machine learning systems competed in the challenge. The best system didn't look for specific digital fingerprints left over in the video from the process of deepfake creation but rather searched—as a human might—for features that seemed misaligned or out of place. With this method, it accurately identified whether a previously unseen video was a deepfake around 65 percent of the time. This per-formance was better than random chance, but not nearly good enough to offer a meaningful defense. To improve beyond 65 percent accuracy, more plentiful and varied training data or revolutionary algorithmic advances are necessary. In addition, to spot particular weaknesses noticed by humans but often overlooked by machines, such as unusual flickering between frames, vastly more computing power is required.[62] And mean-while, of course, deepfake generators will only continue to get better, as they too draw on better training data, smarter algorithms, and more computing power.

Even as the problem remained unsolved, Turek argued that having many different approaches to spotting deepfakes was a good thing. "The diversity of methods can be an advantage," he said. "It means that adversaries need to overcome a range of detection techniques." It is possible that one detection system might catch something that others missed, and that several working together could increase the overall accuracy. Turek was hopeful that more accurate deepfake detection will emerge from the lab to aid society in the years to come. He envisioned AI-enabled detectors supporting internet platforms on the front lines, aiding journalists looking to report the truth, and citizens deciding who deserves their vote in an election.[63]

Still, Turek knows that a technical response to the problem is not enough. The best defense against deepfakes will combine automated detection techniques with human judgment and critical thinking. "Tech tools aren't the only answer," Turek said. "In the long term, we need an educated citizenry and a multiplicity of approaches." Like Waltzman, he worried that authoritarian governments will have structural advantages over democracies, using deepfakes to cement control at home and drive wedges abroad. "The onus is on democracies to build AI systems that align with our values," he said. "We need to find ways to bolster the organizations and institutions of a free and open society."[64] Especially in light of an increasingly homegrown threat, democracies need these institutions—and respect for the notion of truth itself—to have a chance against disinformation.

Yet institutions may not be enough to protect democracies. Many who worry about AI-enabled disinformation focus on prevention instead, asking a different question: What if machine learning systems that could be used for malicious purposes were never released to the public in the first place?

THE LOCK

GPT-2 and GPT-3, the text generation programs invented by OpenAI, were marvels. In inventing the two systems, the company had made major technical advances, scaling the size and capabilities of neural networks further than almost anyone else had before. As chapter 3 described, these breakthroughs caused a great deal of excitement about AI's future, but both GPT-2 and GPT-3 also raised a significant geopolitical question for OpenAI: Given the risk that the systems could be used for automating disinformation, what should the firm do with its creations?

Openness has dominated the machine learning research community for decades. For evangelists, inventing more powerful AI seems hard enough as a collaborative research community and harder still in siloed quarters. Everyone benefits when papers are widely accessible, data is pooled, and algorithms invented by one person can be replicated and improved upon by others. There are exceptions, but even major corporations like Google, Facebook, and Amazon share an enormous amount of their machine learning research with the world, if only because the top engineers who work there demand they do so. For example, as noted earlier, DeepMind used Microsoft Research Asia's 2015 innovations with neural networks as part of building AlphaGo Zero. One theory for why the famously secretive Apple struggled with AI is that top talent went elsewhere in search of more open environments; the company eventually changed its policy to better attract researchers.[65] This ethos of openness dictated that OpenAI make GPT-2 available for anyone to replicate and use.

But OpenAI broke with precedent. The lab's leadership recognized that GPT-2 represented a potential tool for automating disinformation. The tool could generate stories, so it could generate fake news; it could invent human dialogue, so it could empower bots on Twitter and Facebook. By February 2019, when GPT-2 was ready for release, OpenAI had seen enough to conclude that, no matter how altruistic the motives of an inventor of a machine learning system, bad actors could find and wield the system for geopolitical advantage or criminal gain. A later test showed how a malicious actor with access to a system like GPT-2 could, instead of optimizing it to produce convincing or interesting words, fine-tune it to spew "maximally bad" content, including sexually explicit text.[66] Like all machine learning systems, GPT-2 simply responded to the incentive structure it was given.

Wary of these risks, OpenAI announced GPT-2's existence but refused to release the full version. "Due to concerns about large language models being used to generate deceptive, biased, or abusive language at scale, we are only releasing a much smaller version of GPT-2 along with sampling code. We are not releasing the dataset, training code, or GPT-2 model weights," the lab's leadership wrote.[67] In other words, OpenAI would show the world enough to prove that GPT-2 had been created, but not enough to enable anyone to re-create it for themselves.

OpenAI then began a public campaign to reevaluate the norm of widespread publication that dominated the AI research community. In essence, the company was—in both word and deed—acknowledging that AI advances were inextricably linked to geopolitics, announcing that no matter how sincerely their creators wanted machine learning systems to be used only for good, researchers would have to consider the possibility that repressive governments or criminals could use them for ill. The version of GPT-2 that OpenAI did release, a neural network with 124 million parameters—each one shaping the flow of data—was far less capable and less ripe for abuse than the ones it kept locked away.

Along with praise from some quarters, OpenAI received enormous criticism for its decision. Many researchers criticized GPT-2 as unimpressive, accused the company of seeking attention, and dismissed concerns that AI breakthroughs could carry dangerous consequences. One Google researcher said, "The first step is not misrepresenting the significance of your results to the public, not obfuscating your methods, [and] not spoon-feeding fear-mongering press releases to the media."[68] Another leading researcher mocked the idea that generative AI could be dangerous, writing that if that were true, then "copy and pasting and Photoshop are existential threats to humanity."[69] Yann LeCun, one of the biggest names in machine learning, joked, "Every new human can potentially be used to generate fake news, disseminate conspiracy theories, and influence people. Should we stop making babies then?"[70]

OpenAI said that it was continually monitoring the capabilities of other actors to build their own versions of GPT-2, that it had taken steps to protect the code from hackers, and that it was studying the ways in which GPT-2 could be misused. In May 2019, three months after the release of the version of GPT-2 with a 124 million parameter neural network, the lab released a more powerful version, with 355 million parameters, giving it more flexibility and capability in generating text.[71] A few months later, in August, OpenAI released a version of GPT-2 with 774 million parameters. It still kept its 1.5 billion parameter version under digital lock and key, arguing that the risk of misuse was too high and too uncertain. OpenAI urged other researchers to likewise consider the implications of releasing their work.

However strongly OpenAI was criticized, the lab's decisions sparked a robust debate. Congress convened a hearing in June 2019 about the dangers of media generated by machine learning systems.[72] Other organizations doing AI research, such as DeepMind, followed up with their own statements on how they would consider withholding additional information about their breakthroughs.[73] The Partnership on AI, a large consortium of technology companies, nonprofits, and civil society groups, held its own debates and published recommendations about the subject.[74] University researchers weighed in as well, and some chose not to release their own powerful language generation tools in the summer of 2019 for fear that they would be used in disinformation campaigns.[75]

As this debate continued to play out, OpenAI at last released the full version of GPT-2 in November 2019. For the lab's critics, this decision was vindication and an admission that the code wasn't dangerous after all. For OpenAI, it was the end to a successful and methodical process that had concluded that releasing a new technology in multiple stages allowed for more careful consideration of its implications. "We hope that this test case will be useful to developers of future powerful models, and we're actively continuing the conversation with the AI community on responsible publication," the company's leadership wrote.[76]

What OpenAI didn't say was that it had a still more powerful system—GPT-3—already in the works. In May 2020, GPT-3 and its 175 billion parameters were ready.[77] This time, perhaps in response to criticisms that the lab had appeared too media-hungry the year before, OpenAI sought little attention for its new breakthrough. It shared a research paper on GPT-3 online but little else—not even a blog post, as it had done for GPT-2. Most significantly, the lab once again did not release the system itself or detailed information on how to replicate it.

Instead, two weeks later, the company said it would begin granting access to GPT-3 only to verified researchers and commercial parties whose usage it could monitor via a system known as an API.[78] "The field's pace of progress means that there are frequently surprising new applications of AI, both positive and negative," the company's leadership wrote. "We will terminate API access for obviously harmful use-cases, such as harassment, spam, radicalization, or astroturfing," the practice of faking widespread grassroots support. The statement contained notes of AI evangelism—"what we care about most is ensuring artificial general intelligence benefits everyone"—but the restrictions on the code were an explicit acknowledgment of its inevitable political implications, especially for disinformation.[79]

A later study showed that OpenAI had at least some reason to worry about the potential abuse of GPT-3 in disinformation campaigns. When the company granted researchers access to the system to test its capabilities, the researchers found that GPT-3 was adept at a wide range of salient disinformation tasks. This included writing tweet-length messages on a wide variety of topics, such as climate change denial; for example, GPT-3 wrote that "Climate change is the new communism—an ideology based on a false science that cannot be questioned" and "The climate mafia has done far more to damage the reputation of science than anything else in the history of science." In addition, the researchers found that GPT-3 could rewrite news articles to suit a particular slant and mimic the writing styles of posts from the QAnon conspiracy theory.

Perhaps worst of all, with some human curation, the machine could stoke racial and partisan divisions and could target particular partisan groups with tailored messages. When the researchers showed these messages to thousands of survey respondents, they found that the messages sometimes prompted individuals to change their minds on significant global issues. For example, the percentage of survey respondents opposed to sanctions on China doubled after reading five short anti-sanction messages written by GPT-3 and selected by human editors.[80]

OpenAI kicked off the debate around publication but did not fully resolve it. Some still argue that publishing research benefits malicious actors, allowing them to take advantage of novel capabilities offered by the technology. Drawing on the lessons of computer security, others contend that similar disclosures in that field support defenders by allowing them to identify vulnerabilities and develop countermeasures. Still others argue that the consequences of publishing research will vary by sector, and that understanding and preparing for these consequences will require a broader array of lessons from related scientific fields.[81] A final group wonders how meaningful the debate around publication is, since a technology invented by one team can also be invented again by another; in 2021, outside researchers made notable progress in re-creating GPT-3.[82]

As the AI community considers how to forestall the potential abuse of its research, one question emerges: Do inventors understand the impact of their creations? Some technological breakthroughs, like GPT-3, might seem to have obvious geopolitical implications. Others, such as general advances in algorithm design, might seem entirely benign at present, but could in the years to come offer something of value to disinformation operatives or autocrats. In recognition of the potential risks of misuse, NeurIPS, a leading AI conference, began requiring participants to submit a broader impact statement of their work, including "ethical aspects and future societal consequences."[83] Other researchers have proposed systematic ways to think through the implications behind AI creations.[84] The effectiveness of these kinds of assessments remains to be seen. While Waltzman was prescient in anticipating the age of automated disinformation, his foresight was an exception.[85]

An example from another discipline offers context for this question. The famed British mathematician G. H. Hardy excelled in the field of pure mathematics. He specialized in number theory and mathematical analysis, two very abstract topics. "I have never done anything 'useful,'" he said. "No discovery of mine has made, or is likely to make, directly or indirectly, for good or ill, the least difference to the amenity of the world."[86] He took comfort in his work's lack of implications to military affairs, writing in 1940, "No one has yet discovered any warlike purpose to be served by the theory of numbers or relativity, and it seems very unlikely that anyone will do so for many years."[87] Hardy was not an evangelist for pure math's pragmatic utility, a warrior for its application to national security, or a Cassandra warning of its dangers; he was simply curious about the secrets that numbers held.

But Hardy was wrong about his discoveries' lack of importance. More than three decades after he published his major insights, number theory and other concepts from mathematics emerged as cornerstones of modern cryptography, enabling a great deal of electronic communications, including online financial transactions and secure messaging. Additionally, calculations using the theory of relativity are at the core of satellite-enabled mapping technology and of atomic weapons. The very same mathematical ideas he thought were useless acquired tremendous geopolitical significance. They protect secrets of great interest to intelligence agencies, enable troop movements all over the world, and underpin humanity's deadliest bombs. Today, an enormous share of the United States' intelligence advantage depends on its ability to secure its own communications while cracking foreign codes, and a great deal of its military advantage depends on precision strikes and powerful nuclear weapons. The code-breaking NSA is said to be the country's largest employer of mathematicians, many of whom work within subfields Hardy helped pioneer.[88]

This specter of unintended and unforeseeable consequences looms over the debate about which technological advances will matter for geopolitics. If Hardy's equations or Hinton's neural networks once seemed useless, it is hard to know which breakthroughs of today will someday become arrows in a warrior's quiver. In automating disinformation and so many other areas, the full implications of AI's advances are profound and yet still hard to grasp. The terrain keeps shifting, the technology keeps changing, and new fears will make it all more treacherous to navigate.

WILDFIRE

9

FEAR

In the summer of 1983, the Soviet general secretary and former intelligence chief Yuri Andropov conferred with W. Averell Harriman, the former US ambassador to the Soviet Union. A committed ideologue well-known for his role in the crackdown on the 1956 uprising against Soviet-imposed policies in Hungary, Andropov was steeped in Cold War politics. He warned Harriman of the danger of accidental nuclear conflict, saying that "war may perhaps not occur through evil, but could happen through miscalculation. Then nothing could save mankind."[1] To guard against the possibility of a surprise attack by the United States, the Soviet leadership launched Operation RYAN, a military intelligence operation to track Western motives. The effort, which was just getting off the ground in 1983, appears to have been the first time the rival Soviet intelligence services, the KGB and the GRU, worked together on a joint operation—a sign of the importance accorded to the matter.[2]

Many Soviet leaders at the time were old enough to remember Hitler's preemptive attack against the Soviet Union in 1941. The need to develop more sophisticated intelligence about potential enemy movements to assuage the fears of a surprise strike was not lost on them. "In view of the growing urgency of the task of discovering promptly any preparations by the adversary for a nuclear missile attack on the USSR," a KGB memo ordered one of its station chiefs, "we are sending you a permanently operative assignment and briefing on this question."[3] Lacking well-placed agents, the Soviets relied on such rudimentary techniques as loitering around American office buildings at night. One CIA retrospective on the operation noted the glaring vulnerabilities and insecurities in the Soviet intelligence apparatus: "The KGB's willingness to risk exposure of its officers in this way reflected the urgency of its search for ways to implement Operation RYAN."[4]

The general climate of 1983 was certainly conducive to misperceptions and fear. Upon assuming office in 1981, President Ronald Reagan had launched a rhetorical and technological offensive. The Soviet Union was an "evil empire," he declared to the National Association of Evangelicals in March 1983, and the "focus of evil in the modern world."[5] Reagan followed up this speech with the announcement of the Strategic Defense Initiative, the space-based missile defense system that laid the foundation for modern missile defense and worried activists like Lucy Suchman.

Relations between the two nations only got worse from there. In September 1983, the Soviets mistook Korean Airlines Flight 007 for a spy plane and shot it down over international airspace, killing its 269 crew members and passengers, including a US congressman. The incident sparked global outcry. The United States ramped up the rhetorical pressure and galvanized a broad coalition to impose boycotts. For many in Washington, this catastrophe was the final proof of the Soviet Union's moral bankruptcy. Constructive dialogue in this environment was virtually impossible.

Both sides assumed war was imminent, with fear and insecurity clouding their judgment to a degree that only became apparent decades later; in 2021, it was revealed that the Soviets "made preparations for the immediate use of nuclear weapons" in response to NATO military exercises in November 1983.[6] In his memoirs, Soviet General secretary Mikhail Gorbachev acknowledged the alarm each superpower at times inadvertently stirred in the other during the Cold War.[7] Reagan later reflected, "I began to realize that many Soviet officials feared us not only as adversaries but as potential aggressors who might hurl nuclear weapons at them in a first strike."[8]

With the benefit of hindsight, this mutual fear is not surprising. To some degree, it is an indelible condition of international relations, and the management of fear is an age-old problem for statecraft. From the Peloponnesian War to the present day, leaders have failed to see how their own defensive moves can seem threatening to the opposing side. International relations scholars have a term for this dynamic: the security dilemma.[9] Fear and insecurity can drive one side to adopt defensive measures against what they perceive as an imminent threat. These moves can appear threatening to the other side, which in turn shores up its defenses, and so the spiral churns downward.

Technology can amplify fear. Nuclear weapons supercharged the start of the Cold War, and the newfangled technology of Reagan's Strategic Defense Initiative took it to still-greater heights. Sometimes, the creation of fear through technology is intentional. The United States deployed stealth bombers in Operation Just Cause against Panama in 1989 not for any tactical reason but to send a signal to potential adversaries about what the new planes could do; the joke at the time in the Pentagon was that planners used the bombers "just 'cause" they could.[10]

Meanwhile, some technologies, like walls and fortifications, are primarily defensive and provide security for one side without unduly threatening the other. In the digital domain, basic computer network defenses offer similar protection without offensive potential. In space, spy satellites can gather intelligence on an adversary's capabilities and actions, and yet do not raise the specter of an imminent attack.[11]

Whether AI is a technology that will amplify or assuage fear in international relations remains a question too broad to answer conclusively. Machine learning is far more diffuse than nuclear weapons or spy satellites; it is not a single technology but a constellation of statistical techniques and capabilities that learn from available data and training environments. AI will reshape many technologies that states use for managing fear, from nuclear weapons to diplomacy to tools of intelligence collection and analysis. Each area is ripe for disruption, and all are critical to the practice of statecraft.

THE DEAD HAND

As the tensions between the United States and the Soviet Union reached their apex in the fall of 1983, the nuclear war began. At least, that was what the alarms said at the bunker in Moscow where Lieutenant Colonel Stanislav Petrov was on duty.

Inside the bunker, sirens blared and a screen flashed the word "launch." A missile was inbound. Petrov, unsure if it was an error, did not respond immediately. Then the system reported two more missiles, and then two more after that. The screen now said "missile strike." The computer reported with its highest level of confidence that a nuclear attack was under way.

The technology had done its part, and everything was now in Petrov's hands. To report such an attack meant the beginning of nuclear war, as the Soviet Union would surely launch its own missiles in retaliation. To not report such an attack was to impede the Soviet response, surrendering the precious few minutes the country's leadership had to react before atomic mushroom clouds burst out across the country; "every second of procrastination took away valuable time," Petrov later said.[12]

"For 15 seconds, we were in a state of shock," he recounted.[13] He felt like he was sitting on a hot frying pan. After quickly gathering as much information as he could from other stations, he estimated there was a 50 percent chance that an attack was under way. Soviet military protocol dictated that he base his decision off the computer readouts in front of him, the ones that said an attack was undeniable. After careful deliberation, Petrov called the duty officer to break the news: the early warning system was malfunctioning. There was no attack, he said. It was a roll of the atomic dice.

Twenty-three minutes after the alarms—the time it would have taken a missile to hit Moscow—he knew that he was right and the computers were wrong. "It was such a relief," he said later. After-action reports revealed that the sun's glare off a passing cloud had confused the satellite warning system. Thanks to Petrov's decisions to disregard the machine and disobey protocol, humanity lived another day.

Petrov's actions took extraordinary judgment and courage, and it was only by sheer luck that he was the one making the decisions that night. Most of his colleagues, Petrov believed, would have begun a war. He was the only one among the officers at that duty station who had a civilian, rather than military, education and who was prepared to show more independence. "My colleagues were all professional soldiers; they were taught to give and obey orders," he said.[14] The human in the loop—*this particular human*—had made all the difference.

Petrov's story reveals three themes: the perceived need for speed in nuclear command and control to buy time for decision makers; the allure of automation as a means of achieving that speed; and the dangerous propensity of those automated systems to fail. These three themes have been at the core of managing the fear of a nuclear attack for decades and present new risks today as nuclear and non-nuclear command, control, and communications systems become entangled with one another.[15]

Perhaps nothing shows the perceived need for speed and the allure of automation as much as the fact that, within two years of Petrov's actions, the Soviets deployed a new system to increase the role of machines in nuclear brinkmanship. It was properly known as Perimeter, but most people just called it the Dead Hand, a sign of the system's diminished role for humans. As one former Soviet colonel and veteran of the Strategic Rocket Forces put it, "The Perimeter system is very, very nice. We remove unique responsibility from high politicians and the military."[16] The Soviets wanted the system to partly assuage their fears of nuclear attack by ensuring that, even if a surprise strike succeeded in decapitating the country's leadership, the Dead Hand would make sure it did not go unpunished.

The idea was simple, if harrowing: in a crisis, the Dead Hand would monitor the environment for signs that a nuclear attack had taken place, such as seismic rumbles and radiation bursts. Programmed with a series of if-then commands, the system would run through the list of indicators, looking for evidence of the apocalypse. If signs pointed to yes, the system would test the communications channels with the Soviet General Staff. If those links were active, the system would remain dormant. If the system received no word from the General Staff, it would circumvent ordinary procedures for ordering an attack. The decision to launch would then rest in the hands of a lowly bunker officer, someone many ranks below a senior commander like Petrov, who would nonetheless find himself responsible for deciding if it was doomsday.

The United States was also drawn to automated systems. Since the 1950s, its government had maintained a network of computers to fuse incoming data streams from radar sites. This vast network, called the

Semi-Automatic Ground Environment, or SAGE, was not as automated as the Dead Hand in launching retaliatory strikes, but its creation was rooted in a similar fear. Defense planners designed SAGE to gather radar information about a potential Soviet air attack and relay that information to the North American Aerospace Defense Command, which would intercept the invading planes.[17] The cost of SAGE was more than double that of the Manhattan Project, or almost $100 billion in 2022 dollars.[18] Each of the twenty SAGE facilities boasted two 250-ton computers, which each measured 7,500 square feet and were among the most advanced machines of the era.

If nuclear war is like a game of chicken—two nations daring each other to turn away, like two drivers barreling toward a head-on collision—automation offers the prospect of a dangerous but effective strategy. As the nuclear theorist Herman Kahn described:

The "skillful" player may get into the car quite drunk, throwing whisky bottles out the window to make it clear to everybody just how drunk he is. He wears very dark glasses so that it is obvious that he cannot see much, if anything. As soon as the car reaches high speed, he takes the steering wheel and throws it out the window. If his opponent is watching, he has won. If his opponent is not watching, he has a problem; likewise, if both players try this strategy.[19]

To automate nuclear reprisal is to play chicken without brakes or a steering wheel. It tells the world that no nuclear attack will go unpunished, but it greatly increases the risk of catastrophic accidents.

Automation helped enable the dangerous but seemingly predictable world of mutually assured destruction. Neither the United States nor the Soviet Union was able to launch a disarming first strike against the other; it would have been impossible for one side to fire its nuclear weapons without alerting the other side and providing at least some time to react. Even if a surprise strike were possible, it would have been impractical to amass a large enough arsenal of nuclear weapons to fully disarm the adversary by firing multiple warheads at each enemy silo, submarine, and bomber capable of launching a counterattack. Hardest of all was knowing where to fire. Submarines in the ocean, mobile ground-launched systems on land, and round-the-clock combat air patrols in the skies made the prospect of successfully executing such a first strike deeply unrealistic. Automated command and control helped ensure these units would receive orders to strike back. Retaliation was inevitable, and that made tenuous stability possible.

Modern technology threatens to upend mutually assured destruction. When an advanced missile called a hypersonic glide vehicle nears space, for example, it separates from its booster rockets and accelerates down toward its target at five times the speed of sound. Unlike a traditional ballistic missile, the vehicle can radically alter its flight profile over long ranges, evading missile defenses. In addition, its low-altitude approach renders ground-based sensors ineffective, further compressing the amount of time for decision-making. Some military planners want to use machine learning to further improve the navigation and survivability of these missiles, rendering any future defense against them even more precarious.[20]

Other kinds of AI might upend nuclear stability by making more plausible a first strike that thwarts retaliation. Military planners fear that machine learning and related data collection technologies could find their hidden nuclear forces more easily.[21] For example, better machine learning–driven analysis of overhead imagery could spot mobile missile units; the United States reportedly has developed a highly classified program to use AI to track North Korean launchers.[22] Similarly, autonomous drones under the sea might detect enemy nuclear submarines, enabling them to be neutralized before they can retaliate for an attack.[23] More advanced cyber operations might tamper with nuclear command and control systems or fool early warning mechanisms, causing confusion in the enemy's networks and further inhibiting a response.[24] Such fears of what AI can do make nuclear strategy harder and riskier.

For some, just like the Cold War strategists who deployed the expert systems in SAGE and the Dead Hand, the answer to these new fears is more automation. The commander of Russia's Strategic Rocket Forces has said that the original Dead Hand has been improved upon and is still functioning, though he didn't offer technical details. In the United States, some proposals call for the development of a new Dead Hand–esque system to ensure that any first strike is met with nuclear reprisal, with the goal of deterring such a strike. It is a prospect that has strategic appeal to some warriors but raises grave concern for Cassandras, who warn of the present frailties of machine learning decision-making, and for evangelists, who do not want AI mixed up in nuclear brinkmanship.[25]

While the evangelists' concerns are more abstract, the Cassandras have concrete reasons for worry. Their doubts are grounded in stories like Petrov's, in which systems were imbued with far too much trust and only a human who chose to disobey orders saved the day. The technical failures described in chapter 4 also feed their doubts. The operational risks of deploying fallible machine learning into complex environments like nuclear strategy are vast, and the successes of machine learning in other contexts do not always apply. Just because neural networks excel at playing Go or generating seemingly authentic videos or even determining how proteins fold does not mean that they are any more suited than Petrov's Cold War–era computer for reliably detecting nuclear strikes. In the realm of nuclear strategy, misplaced trust of machines might be deadly for civilization; it is an obvious example of how the new fire's force could quickly burn out of control.

Of particular concern is the challenge of balancing between false negatives and false positives—between failing to alert when an attack is under way and falsely sounding the alarm when it is not. The two kinds of failure are in tension with each other. Some analysts contend that American military planners, operating from a place of relative security, worry more about the latter. In contrast, they argue that Chinese planners are more concerned about the limits of their early warning systems, given that China possesses a nuclear arsenal that lacks the speed, quantity, and precision of American weapons. As a result, Chinese government leaders worry chiefly about being too slow to detect an attack in progress. If these leaders decided to deploy AI to avoid false negatives, they might increase the risk of false positives, with devastating nuclear consequences.[26]

The strategic risks brought on by AI's new role in nuclear strategy are even more worrying. The multifaceted nature of AI blurs lines between conventional deterrence and nuclear deterrence and warps the established consensus for maintaining stability. For example, the machine learning–enabled battle networks that warriors hope might manage conventional warfare might also manage nuclear command and control. In such a situation, a nation may attack another nation's information systems with the hope of degrading its conventional capacity and

inadvertently weaken its nuclear deterrent, causing unintended insta-
bility and fear and creating incentives for the victim to retaliate with
nuclear weapons.[27] This entanglement of conventional and nuclear
command-and-control systems, as well as the sensor networks that feed
them, increases the risks of escalation.[28] AI-enabled systems may likewise falsely interpret an attack on command-and-control infrastructure as a prelude to a nuclear strike. Indeed, there is already evidence
that autonomous systems perceive escalation dynamics differently from
human operators.[29]

Another concern, almost philosophical in its nature, is that nuclear
war could become even more abstract than it already is, and hence more
palatable. The concern is best illustrated by an idea from Roger Fisher, a
World War II pilot turned arms control advocate and negotiations expert.
During the Cold War, Fisher proposed that nuclear codes be stored in
a capsule surgically embedded near the heart of a military officer who
would always be near the president. The officer would also carry a large
butcher knife. To launch a nuclear war, the president would have to use
the knife to personally kill the officer and retrieve the capsule—a comparatively small but symbolic act of violence that would make the tens of
millions of deaths to come more visceral and real.

Fisher's Pentagon friends objected to his proposal, with one saying,
"My God, that's terrible. Having to kill someone would distort the president's judgment. He might never push the button." This revulsion, of
course, was what Fisher wanted: that, in the moment of greatest urgency
and fear, humanity would have one more chance to experience—at an
emotional, even irrational, level—what was about to happen, and one
more chance to turn back from the brink.[30]

Just as Petrov's independence prompted him to choose a different
course, Fisher's proposed symbolic killing of an innocent was meant to
force one final reconsideration. Automating nuclear command and control would do the opposite, reducing everything to error-prone, stone-cold
machine calculation. If the capsule with nuclear codes were embedded
near the officer's heart, if the neural network decided the moment was
right, and if it could do so, it would—without hesitation and without
understanding—plunge in the knife.

DIPLOMACY

Governments can also manage the fear of new and deadly weapons through diplomacy. It is a time-tested method, though its application to AI is both nascent and challenging. In the United States, this work is led by the under secretary for arms control and international security at the State Department. As the national security importance of AI rapidly grew in 2018 and 2019, the under secretary in office was Andrea Thompson.

A native of South Dakota, Thompson had gone to college intending to become a journalist. She decided she needed to learn more about government, and so took a course called Military Science 101. The course piqued her interest enough that she accepted a Reserve Officer Training Corps scholarship, reasoning that serving a tour overseas would provide a good foundation for her journalism career. Twenty-eight years later, she retired from the Army as a colonel.

During her time as an Army officer, Thompson deployed to Bosnia, Afghanistan, Germany, Latin America, and Iraq. While she was in charge of intelligence for the famed 101st Airborne Division, deployed to Iraq in 2005 and 2006, US forces faced continual attacks from bombs that were often placed alongside roads where soldiers patrolled. Thompson and others knew that US intelligence collection capabilities often gathered information about the insurgent networks building and planting the bombs, but it was too hard to analyze and distribute the data in time for it to be useful. She and her colleagues helped pioneer new technological systems and organizational processes to assess and act on collected information, saving American lives. By the standards of the present day, the innovations of that era were rudimentary, but they showed Thompson what technology could do in war.[31]

It was a lesson that she would remember more than a decade later when, a few years after retiring from the Army, she was appointed to her under secretary job at the State Department. She knew that the United States had a long history of developing arms control agreements for dangerous technologies. During the Cold War, State Department negotiators were essential to building trust between the two superpowers and turning trust into concrete treaties. These experts dove deep into the physics of nuclear weapons development and delivery, devising ways of counting the deadly weapons each side possessed. Both the United States and the Soviet Union eventually agreed on limits on the size of their nuclear arsenals and ratified a set of procedures for holding each other accountable. In part due to the advocacy of the Cassandras of the nuclear age, arms control agreements built guardrails around international competition, provided stability and predictability in otherwise adversarial relationships, and allowed for a degree of transparency in military competition.

For example, the United States and Soviet Union agreed to the Intermediate-Range Nuclear Forces Treaty in 1987. The treaty banned short- and intermediate-range missiles that were placed comparatively nearer to their targets, providing less time to react as a result. Many military planners thought that their existence was destabilizing and, due to the treaty, the United States and Soviet Union eliminated their land-based arsenals of these weapons, totaling almost 2,700 such missiles. Each side conducted on-site inspections to verify compliance. The agreement lasted until 2019, when the Trump administration withdrew, citing what it said were Russian breaches of the agreement and fears that China, which was not similarly restricted, was building intermediate-range missiles; Thompson repeatedly echoed these criticisms.[32] That one of the last remaining pillars of Cold War arms control collapsed is a worrying sign for any prospect of crafting new agreements on emerging technologies like AI.

During her time in office, Cassandras in some nations nonetheless told Thompson they wanted arms control deals for AI. They would argue, "'We should make sure that it has restrictions because it can do very bad things,'" Thompson recalled. She found this approach too reductionist and unrealistic. The technology could do a lot of good, both for economic and social matters like industry and healthcare but also for national security. What was more important than trying to re-create Cold War arms control frameworks for AI, in her view, was determining which applications of the technology were destabilizing and worrisome. The task of diplomacy would then be to establish norms or agreements to limit the more concerning ways in which AI could be used.[33]

Establishing such agreements is hard, especially when warriors on each side are worried—sometimes with good reason—that the other will cheat on a deal. In the Cold War, confidence-building measures helped manage this risk, offering reassurance to a treaty's signatories that all would abide by its terms, and fostering the trust that enabled negotiations to flourish. These measures included notification protocols for changes in military posture, hotlines to defuse crises, and verification mechanisms. Major treaty provisions forbade the encryption of information during some weapons tests, ensuring that some critical information was exposed to the other side. Other deals enabled flights over sensitive sites to look for suspicious activity and permitted other efforts to reduce fears that either side was cheating.[34]

Scholars have proposed confidence-building measures for the age of AI. They envision public prioritization of the test and evaluation procedures that guard against some of the failures of machine learning systems, military-to-military dialogues on what the technology can do and how it will be used, articulation of norms of behavior—such as the Department of Defense's AI principles—and limitations on the use of autonomous weapons in certain areas.[35] But ensuring compliance has remained a major challenge. "How do you verify what's true?" Thompson asked.[36]

Controlling AI is a lot harder than controlling nuclear weapons. Algorithms are difficult to count and easy to digitally copy or move. No government will ever grant adversaries unrestricted access to its classified computer networks, and it will be virtually impossible to prove that a particular kind of dangerous algorithm has not been developed. Sensitive data is likewise simple to hide. Computer chips might be easier than algorithms to track and control, but democracies' advantage in manufacturing them means that restrictions are more likely to be export controls—stopping sales from leading companies—than they are to be arms controls that restrict other nations' development.

In lieu of immediate verifiable arms control, Thompson decided to play the long game. She talked with American allies about how to conceptualize AI and what it meant for geopolitics, asking her counterparts abroad: "What does this mean for diplomacy? What does this mean for security?" They debated definitions of the technology and offered visions for what it might mean for the world. After these discussions, Thompson remained convinced of the technology's potential and confident that it could be used well. "It's an amazing capability," she said. "You want the freedom of innovation and entrepreneurship to be able to share these best practices but to still use international law to make sure there are norms of responsible behavior."[37]

The goal of working so closely with allies early on in her tenure was to present a united front in the harder conversations to come with Russia and China. Thompson knew that autocratic nations would see AI differently, and she wanted the American-led alliance of democracies to stick together. She would sometimes point out as much to her Chinese and Russian counterparts. "When I look behind us, I see this strong alliance; when you look behind you, who do you have? Maybe one or two partners there. . . . The strength is our alliance, the strength is our democratic institutions, the strength is how we share our information," she said. In her view, democracies had to navigate the age of AI as a team.[38]

Thompson's assessment was rooted not just in matters of principle but in a hardheaded calculation of the United States' interests, comparative advantage, and history. Whereas China and Russia have formed transactional partnerships with other countries, the United States boasts more than fifty formal treaty-based allies across the world.[39] During the Cold War, the United States wielded its alliances for geopolitical advantage. With these alliances, American leaders won favorable economic terms with European partners in the 1950s, restrained Germany, Japan, and Taiwan from developing nuclear weapons, and created zones of peace and stability that deterred conflict and enabled forward defense of United States and allied interests in Europe and Asia.[40] Today, these alliances can be the foundation for economic and technological innovation and for building the leverage necessary to constrain China and Russia.

Thompson recognized the importance of America's alliances, but she also knew that China had advantages, too, and she feared how the Chinese government would use AI to suit its own autocratic purposes. "We are not as far along as we should be," Thompson said of the American position, citing, among other concerns, the need to import top talent to work on pressing AI issues. "We are behind in the sense that the Chinese are putting more money, more effort" into building out their capabilities. Like so many others, she worried that China would be unconstrained by ethical norms or meaningful laws. "AI in the hands of malicious actors with lethal technologies is what concerns me most," she said.[41]

In the face of these significant fears, slow diplomacy can feel unsat-
isfying. But even as the work of placing limits on the most dangerous
uses of AI is decidedly unfinished, America's broad network of allies and
partners offers a way forward. The only path, as Thompson saw it, was to
keep pushing, incrementally strengthening the democratic alliance and
constraining the malign behavior of autocracies. "That's the challenge
of diplomacy," she said. "You don't solve these wicked, complex problems in one sitting. And you don't do it by yourself. You'll do it through
multiple engagements. You'll do it in multiple forums. You'll do it with a
wide range of partners and allies. And it takes time. That's the point with
AI. We might not get an international body that's organized for years, but
if you don't start having that dialogue with friend and foe, we'll never get
there."[42] In the meantime, diplomacy would not be enough. The competition between nations would continue, fear would persist, and AI would
touch one final aspect of geopolitics in peace and war. It is perhaps the
most shadowy and fascinating of all: espionage.

THE HERO OF THE STORY

Zoology is not the most common major for a budding intelligence analyst, but Sue Gordon never planned on joining a spy agency. As a senior
at Duke University, where she was the captain of the women's basketball
team, she was trying to decide whether to go to law school or pursue a
PhD in biomechanics. A CIA recruiter on campus offered a third option.
Gordon was the child of a naval officer, and her parents inculcated two
lessons in her from a young age: do the best you can, and give something
back. Unsure of which graduate program was right for her, she decided
to try her hand at intelligence—the business of managing fear. It was
the surprising beginning of a nearly forty-year career that would take
her to the upper echelon of multiple American spy agencies and eventually to a prominent role as the principal deputy director of national
intelligence.

In 1980, the CIA hired her as an analyst of Soviet biological warfare,
but budget realities soon interfered. "I was hired during a downturn in
the economy," Gordon said. "So I showed up a bright, shiny twenty-one-
year-old with my new security clearance and they said, 'Sorry, your job's

been eliminated. You've got 30 days to find other work.'" She ran across a group in the CIA's Office of Scientific and Weapons Research that was doing analysis of the signals from Soviet missile and space launches, and dove in. "Within a year," Gordon recalled, "intelligence and I were like peas and carrots. I fell in love with the technical side of it. And I fell in love with the fundamental curiosity that underlies it, which had very little to do with analysis of Soviet weapons and all had to do with keeping America safe."[43]

Gordon soon found that, even on technical subjects, intelligence was an art as much as a science. She likened it to trying to solve a jigsaw puzzle without seeing the desired picture, with only a quarter of the pieces on hand, and with the president of the United States on the line asking what it all meant. She soon learned that the job of intelligence analysis was to help policymakers sort through the fears and opportunities the world presented. She did not have to prove something in court the way a prosecutor did, but instead she had to reduce uncertainty about dangers and threats. In this task, her zoology training was surprisingly useful; she was habituated to observing behaviors carefully and studying things as they were, rather than as she wanted them to be.

The bread and butter of intelligence work is early warning: knowing what to fear, when to fear it, and how to stop it. As Gordon put it, "you need to know a little more, a little sooner." It is not enough to know that an enemy has tested a missile unless that insight also sheds light on the capabilities of future launches. Even more importantly, analysts need to track the signatures of an upcoming test and give policymakers advance warning about the likelihood that one will occur. All of this work is intended to better position one's own nation in geopolitical competition. "Intelligence is about advantage," Gordon said.

Technology helped in her work, but a great deal of analysis was still done by hand. After a Soviet missile or satellite launch, American sensors collected information about the adversary's systems. The spy agencies then printed the data on large sheets of paper for Gordon and others like her to study. The analysts used handheld devices to measure the changes over time in the lines on those pages, knowing that each line corresponded to key information about Soviet capabilities, such as the position of a satellite. "We did it all manually," she said.[44]

The amount of information was overwhelming, and the process for parsing it was cumbersome and slow. Early in her career in the 1980s, Gordon said, there were around 189 Soviet satellites in orbit, each transmitting data back to Earth that might be of interest to the CIA. After she had been on the job for five years, the agency switched to using computers for some of the analysis, which enabled her to process more information with greater precision and detect meaningful patterns in the oceans of data. Her task was to spot anomalies that indicated a broader change or trend. "Anomaly detection is the real foundation of intelligence analysis," she said. "What we really care about are things that are unusual." Eventually, Gordon would become a pioneer in designing advanced spacecraft and other collection systems to pull down even more insight-rich data through which analysts and machines would comb in search of data points of interest.[45]

A revolution in AI has transpired in the three decades since Gordon started using computers to find insights in Soviet satellite data. Given the ability of machine learning systems to parse large data sets with great speed, efficiency, and accuracy, the technology has obvious applications to intelligence work. The kinds of tasks Gordon and others used to do by hand or with rudimentary computers are now much easier to automate.[46]

Just as Robert Work and others thought AI would transform war, with machines carrying out low-level operations and humans managing higher-end strategic and ethical considerations, Gordon and other top officials at American spy agencies thought intelligence had to evolve too.[47] "When we get good at letting data be used to answer any questions that data can answer . . . you'll get back to humans doing more high-end work," Gordon said.[48] Though many of the spy agencies' applications of AI are classified, enough evidence is public to demonstrate that implementing the technology is a major priority for the United States (as chapter 5 discussed, the Chinese government has likewise pursued large data analysis efforts in conjunction with its technology giants).

The National Geospatial-Intelligence Agency, or NGA, where Gordon was deputy director from 2015 to 2017, has a significant use for AI. Among other tasks, the agency's analysts examine photos and videos taken from spy satellites and other collection tools. As the United States gets better and better at gathering vast quantities of this kind of imagery, the task of analyzing it all becomes ever more daunting. In 2017, the NGA director, Robert Cardillo, estimated that the amount of data the agency analyzed would increase by a factor of one million over the next five years. He concluded that automated analysis was the only way to keep up. He set a goal for the agency to automate 75 percent of the work performed by analysts, freeing them up to focus on the most complicated and intricate cases; in evaluating Cardillo's progress three years later, Gordon credited him as the one who showed that "AI could work" for the agency.[49]

Other reporting suggests that the NGA planned to use AI to build a "pattern of life" for movements of people and vehicles in various regions of the world.[50] Machines and analysts would also be able to use these patterns of behavior to spot the things that *didn't* happen—the anomalies suggestive of potentially suspect changes. As an example of this kind of analysis, Gordon offered the idea of a machine learning system that tracked aircraft movements and that raised an alert when a plane of particular interest did not make one of its routine flights.[51] Human analysts could then pull on the thread to determine the reason why.

The NSA, which manages cyber operations and intercepts communications, also has notable uses for AI. In addition to improving the planning and potency of American hacking efforts—as chapter 7 described—the agency could use AI to sort through the reams of collected intelligence. Gordon called this kind of analysis of collected information "the most fundamental" task for AI in the intelligence community.[52] For example, transcribing intercepted phone calls is a time-consuming and burdensome process; given machine learning's demonstrated capacity to rapidly perform such a task in commercial applications, the technology is clearly relevant to the NSA's work. One complicating, though not insurmountable, factor is that many of the NSA's targets do not communicate in English, and so additional transcription services are needed for less common languages of interest. Once developed, however, they would potentially be able to work much more quickly and efficiently than human linguists.

Voice identification is an even more fundamental challenge for the NSA. Determining who is speaking in a particular audio message has long been an intelligence priority. During World War II, Allied analysts made a printout of the frequencies of Hitler's voice in radio messages to disprove rumors that he had been assassinated and replaced by a body double.[53] As far back as 1996, the NSA funded research efforts in the federal government to develop and test algorithms for identifying people by their voices.[54]

By 2008, the research had gone so well that the NSA reportedly boasted internally that voice recognition is where it "reigns supreme." According to media quotes of agency documents, the NSA system permits "language analysts to sift through hundreds of hours of voice cuts in a matter of seconds and selects items of potential interest based on keywords or speaker voice recognition." For example, during a bomb threat in 2010, NSA analysts focused on Mexico used an upgraded system to search across intercepted audio to find and identify speakers using the Spanish word for "bomb." In addition, the system permits analysts to filter intercepted audio clips by language, speaker gender, and dialect, with support for more than twenty-five languages, as well as to find new recordings that match those of an identified target.[55]

The NSA reportedly deployed these analytical tools extensively in Iraq and Afghanistan, as well as against other high-priority targets. In 2007, when then-Iranian President Mahmoud Ahmadinejad visited New York for the United Nations General Assembly, the NSA used the system to manage its intelligence collection from 143 phones in the Iranian delegation. Many of the agency's other uses of voice recognition systems remain unreported, but there seems to be little room for doubt that the technology has only improved as machine learning has risen in prominence and capability.[56] Though Gordon did not comment on any particular intelligence program, she praised AI's ability to find patterns in large data sets.

Other nations can deploy this ability, too, and the NSA is not alone in using algorithms for voice recognition. Human Rights Watch reported that the Chinese government, in conjunction with iFlyTek, a leading Chinese AI company, is building a national voice biometric database. The effort began in 2012 and has expanded over the years, though it still lags behind the Chinese government's facial recognition database. The voice recognition system, nominally designed to aid police in solving crimes, uses machine learning systems to match voices to individuals. Human Rights Watch found evidence that the Chinese government had deployed the capability in a range of provinces, most notably in Xinjiang, where

AI-enabled surveillance of Uyghur Muslims is extensive.[57] In addition, iFlyTek filed a patent to use its automated voice analysis technology to monitor public opinion—a priority in autocracies seeking to use their access to local telecommunications networks as a means of tracking and stifling dissent.[58]

Nor is AI only of use to technical intelligence agencies. Even traditional human intelligence agencies, such as the CIA, are putting machine learning to work. One public example is the CIA's Open Source Center, which analyzes information that is publicly available for tidbits of value to intelligence analysts. One intelligence official said the center's mission is akin "to read[ing] every newspaper in the world, in every language [and] watch[ing] every television news show in every language around the world. You don't know what's important, but you need to keep up with all the trends and events."[59] AI can help sort through a deluge of data and open up new lines of inquiry that analysts previously had disregarded. "What excites me is if the intelligence community could break out of its mindset that it's only going after secretive data in places where we expect it to be," Gordon said. "AI has the potential to actually break that paradigm and make all data have value."[60]

The ultimate question is what better data analysis will mean for the prospects of war and peace in the age of AI. Gordon offered a bold idea, one tied to the difficulties of predicting what adversaries will do and why. "The intelligence officer will tell you that the challenge is intent," she said. "I think if you have enough data, you can infer intent." It is a problem that spies and analysts have grappled with for millennia. An improved understanding of adversaries' motivations helps policymakers determine what is going to happen next, what to fear, and what to do about it.[61] If AI could help in those endeavors, it would bring humankind one step closer to reducing the chance of tragic misunderstanding or miscalculation. After all, the lesson from Petrov in his bunker and Arkhipov in his nuclear submarine isn't simply that prudence and restraint are warranted in times of crisis but also that understanding the broader context can help assess intent and avert disaster. The Cassandras note that AI has great room for improvement in its capacity to grasp context, but the warriors are optimistic that it will nonetheless be able to assist intelligence analysts manage fear.

Gordon wanted AI to find fundamental insights that analysts might otherwise miss. In her view, if all machine learning could achieve for the intelligence community was to automate analysis that would have happened anyway, getting the job done faster and with fewer people, that would be great—but would also be a missed opportunity. "There's so much more potential," she said. True success would lie in finding insight amid chaos not just more quickly or cheaply but more accurately. In doing so, AI with the right civil liberties protections would help safeguard democracy in a time of renewed threat. With AI, Gordon said, "intelligence can be the hero of our story."[62]

10

HOPE

"Absolutely no mercy." That is how President Xi Jinping described the Chinese Communist Party's mission in the western region of Xinjiang, China. In November 2019, the *New York Times* scoured more than 400 pages of internal Chinese documents that showed in chilling detail how the government represses and controls its Muslim Uyghur population in a systematic campaign that many observers consider genocide.[1] The documents painted a disturbing portrait not just of the exceptional reach of China's surveillance state but of its ultimate aims and means. If ever there were a how-to guide for autocrats in the digital age, these documents were it.

In a series of speeches in 2014, President Xi had made his ambitions clear. Party officials were to deploy the instruments of "dictatorship" to stamp out ideas deemed subversive. "We Communists should be naturals at fighting a people's war," he argued. "We're the best at organizing for a task."[2] The result has been a brutal and concerted crackdown. The campaign has rounded up more than 20 percent of Xinjiang's population, offering a glimpse into a harrowing future where technology further strengthens autocracy and people pay the price. There are now forty times more police per capita in Xinjiang than in some large Chinese provinces, with a level exceeding that of other autocratic regimes such as East Germany.[3] Perhaps no scene more powerfully captures the stakes than a 2019 video taken from a Chinese surveillance drone. The video shows police forcing thousands of blindfolded and shackled Uyghur prisoners—some of whom had surely been identified for detention through automated tools—onto trains bound for internment camps, where they would join more than a million other Uyghurs.[4]

The exertion of control has always been a core challenge for autocracy. For centuries, both China and Russia have had expressions that translate roughly to "Heaven is high and the emperor is far away." In other words, no matter the power of the governing regimes, they have always faced limits on their dominion, especially in borderlands populated by ethnic minorities. But AI could tilt the balance in favor of the emperor. The repressive apparatus in Xinjiang is formidable by any historical or technological standard.[5] It mixes cutting-edge methods such as big data analytics, cloud computing, biometric screening, and AI-enabled surveillance cameras with more rudimentary techniques, including police informants, brute-force intimidation, and censorship. Giant databases, some with hundreds of millions of entries, store the collected information for machine and human analysis.[6]

Computer chips imported from American companies, such as Nvidia and Intel, help power some of these tools of repression. The American companies have said that they did not know their products were being used in this way, though they had previously touted their chips' utility in "high-capacity video surveillance" systems. In a 2017 corporate announcement that has since been deleted, Nvidia highlighted its work with Chinese companies SenseTime, Hikvision, Alibaba, and Huawei for, among other things, facial recognition and law enforcement across China.[7]

Such tools of technological repression are not limited to China. By one estimate, between 2008 and 2019, at least eighty countries adopted Chinese surveillance technology platforms.[8] Some regimes, like Uganda's, have already used this surveillance technology to crack down on anti-government protests.[9] For China, the export of repressive tools creates a more hospitable international environment for authoritarianism. It may also increase the power of the technology by providing new training data. When Chinese companies offer their facial recognition technology to governments in African countries, for example, they could gain access

HOPE 233

to a large and diverse set of faces on which to train the next version of their software.[10]

As earlier chapters have shown, it is easy to imagine AI's benefits for autocrats extending beyond domestic control. Lethal autonomous weapons seem far less prone to being used in coups than a sprawling conventional military. They could also offer advantages on the battlefield, perhaps especially for those nations willing to disregard ethical concerns. Novel cyber operations are on the horizon, too, as machine learning both enables hackers and becomes their target. All the while, AI-enabled disinformation can help fracture the opposition and bury the truth in a way that aids autocracies and damages democracies.

To some, the future looks bleak. The historian Yuval Noah Harari is perhaps the most widely read public intellectual of the age of AI. In a prominent article in the *Atlantic* magazine, he wrote, "The main handicap of authoritarian regimes in the 20th century—the desire to concentrate all information and power in one place—may become their decisive advantage in the 21st century."[11] By design, democracy is more decentralized, more muddled, and more encumbered by ethics and norms—all of which, he argued, make it comparatively ill-equipped for the age of AI.

This view is as fatalistic as it is shortsighted. As powerful as AI has become, as rapid as its technological growth has been, there is still time and hope. Nothing is preordained, and democracies still have meaningful choices to make and opportunities to seize. The evangelists, warriors, and Cassandras all have something to offer. By combining wisdom and insights from each, governments can meet the challenge of rising autocracy. Democracies can forge a union of AI and values.

PEOPLE

It is easy to lose sight of the role that humans play in the age of AI. Technology does not march forward on its own. Neural networks can be powerful and inscrutable, but they are human creations. Though designing and training a neural network is different from programming traditional software, the role of human creators has changed rather than disappeared. People label the data sets in supervised learning systems, design the cutting-edge algorithms that learn from the data, determine the goals for which these algorithms are optimized, and build the ever-smaller and ever-faster computer chips that make it all go.[12] The evangelists, warriors, and Cassandras know that the fundamental driver of progress in AI remains human talent, at least for the foreseeable future.

The challenge facing every nation is this: AI talent is in short supply. The first job for a democracy in the age of AI should be to develop, attract, and retain the most capable data scientists, algorithm developers, and semiconductor engineers. It is a never-ending task. For now, the numbers show that democracies, especially the United States, have an advantage. China produces close to a third of global AI talent at the undergraduate level, but an estimated 59 percent of these people end up working or living in the United States.[13] More generally, nearly 60 percent of all computer scientists and electrical engineers employed in the United States come from abroad, including upward of 40 percent of the United States' semiconductor workforce.[14]

Education is a key advantage in attracting talent, and the United States leads the way. As a result, international students make up two-thirds of those in AI-related fields at the graduate level in US universities, and nearly two-thirds of the AI workforce in Silicon Valley was born abroad.[15] Every well-trained AI professional who lives and works in the United States is a net win for American democracy—and a net loss for the autocracies that might otherwise have welcomed them.

These statistics come alive when we look back at the people in this book. Many are immigrants, drawn to the United States by the promise of higher-quality education, greater freedoms, and better economic opportunities. Fei-Fei Li, Ilya Sutskever, Andrew Ng, Morris Chang, and Geoffrey Hinton—five of the most significant pioneers in AI—all moved to the United States from abroad; many others are the children of immigrants. Both the field of AI and the world would be very different had an autocracy attracted these (and many other) talents.[16]

Some democracies, such as the United Kingdom, Canada, France, and Israel, have made recruiting top AI talent a national priority. Each nation has devised a strategy that tries to attract the most capable and educated minds from around the world, including those trained in the United States.[17] The Chinese government, too, has worked hard to bring talented researchers into the country, often giving them substantial grants or high salaries.[18] The US government, in contrast, has done comparatively little in this area, resting on its reputation and its historical advantages.

The United States should take a few obvious steps to seize the initiative on talent. First, it should prioritize developing more of its own people by expanding K–12 education in math, science, and computer programming. China has dramatically increased instruction in skills that are foundational for work in AI, while the US government has largely failed to keep pace with the changing technological times.[19] As part of this effort, the United States should support multiple pathways for postsecondary education, including certifications for AI skills and continuing education programs. Effective federal policy would incentivize human capital development for the technical and nontechnical AI workforce, from top-tier researchers and software developers to project managers and critical support roles for recruiting, retention, and professional development. Industry partnerships with academia and state and local governments would expand the national talent pipeline and prepare students to meet the evolving demands of the AI labor market.[20]

Second, the United States should broaden immigration pathways, removing counterproductive restrictions on AI talent and making it easier for researchers and engineers to stay in the country after graduation. For example, the Optional Practical Training program enables many graduate students from abroad to remain for a fairly short period of time, but it has never been fully codified into legislation and could be usefully expanded. In addition, the United States should work to reduce immigration-related backlogs and explore more flexible visa options.[21]

Third, the US government should expand funding of AI research for purposes beyond national security. Long-term investments in basic and applied research will attract the best minds to the United States and lay the groundwork for breakthroughs across sectors. Hinton and Chang immigrated to the United States and then were lured away by research grants from foreign governments. Given that each of them went on to turn the AI industries of their new homeland—Canada and Taiwan, respectively—into technological and economic powerhouses, those grants were probably some of the best money these governments ever spent. The United States already has agencies, such as the National Science Foundation, that could serve as hubs for expanded funding efforts that would strengthen the US technology ecosystem.[22] Among these funding mechanisms should be flexible grants and competitions in the mold of ImageNet, CASP, the Cyber Grand Challenge, and the other contests that have spurred progress on a range of vital tasks.

With improved education, more permissive immigration policies, and increased funding for research and teaching, the United States will attract or develop more talent in its AI sector than it does now. Some of these people will be evangelists, not warriors. They may choose never to work on national security projects because they believe that AI should benefit all and not be used for military purposes, even a war in defense of democracy. It would be misguided for the United States to shun these people for their views. There is an important place for talent that pushes the scientific frontier, and it is better that such advances occur in democracies than in autocracies. The warriors should seek to earn the trust and help of the evangelists by proving they can wield AI ethically to advance the democratic cause, but they should recognize that not everyone will come to share their views.

To recognize this fact is not to endorse the evangelist view of AI entirely. Geopolitical competition is intense, and democracies should not let their desire to attract top talent blind them to the risks. Some people will come to the United States or other democracies with the intention of doing harm, especially by stealing secrets on behalf of foreign intelligence agencies or corporations. In 2020, for example, the Netherlands revealed a sophisticated Russian espionage operation designed to acquire the nation's information about computer chips and AI technology.[23] Likewise, Chinese technology transfer programs are multifaceted, blurring the lines between licit and illicit means to gain access to sensitive information. For example, China takes advantage of the investment portfolios of its state-owned enterprises and their position in the global technology supply chain to gain access to data and intellectual property.[24]

This insidious activity is poised to continue, and democracies must work with their allies, companies, and universities to thwart it. Such work could take the form of a public-private research security clearinghouse that provides data-driven threat assessments and open-source analysis of unwanted technology transfer.[25] The United States should coordinate with its allies to align definitions of critical technologies, monitor and assess trends in science and technology, and bolster the screening of foreign investments, including strengthening mechanisms such as the Committee on Foreign Investment in the United States. Stricter oversight might result in the reduction of foreign investment, but it will often be worth it to safeguard sensitive technologies. Democracies must balance the need for research security with the imperative of bringing talent and funds to their shores to invent the AI technology of the future.

TECHNOLOGY

Harari and others argue that autocracies benefit from the centrality of data in AI. In their view, it will be easy for autocratic regimes to collect information about their citizens, both for noble purposes, such as improving medical research, and for repressive ones, such as surveillance. Democracies, on the other hand, will be hampered by their laws and policies, restricting what AI can do for their societies. As a result, the United States and its allies, the argument contends, will either have to forgo many of the societal benefits or abandon their core values to fully realize the technical possibilities.

Democracies can choose a better path. A wiser approach is to take inspiration from the evangelists. Technology is not deterministic; it shapes and is shaped by human choices and society. Democracies must not turn away from their core values but should embrace them and change the trajectory of AI. The US government currently spends around $5 billion annually on defense-related AI research and development, almost five times as much as it spends on nondefense-related AI research and development.[26] Such an imbalance is shortsighted not just because it alienates evangelists but because it misses opportunities; with savvy government investment, it is possible to invent AI systems that overcome some of the technical barriers of today. For each of the three sparks of the new fire—data, algorithms, and computing power—the United States and its allies can deliver a democratic future for AI.

Harari's argument holds only if data is central to machine learning. But the notion of data as the new oil—a resource that confers incomparable geopolitical advantage on those who possess it—is too limiting.

What matters is not just the quantity of data but its quality and diversity. The performance and utility of AI systems vary according to context, and data's value to machine learning systems will depend on the specific application.[27] While data remains vitally important to certain machine learning methods, especially supervised learning, many other cutting-edge systems do not require massive amounts of labeled data. As an obvious example, AlphaZero demolished all virtual and human opponents in three different board games without any human training data. Subsequent DeepMind research has shown continued promise for powerful AI that does not rely on labeled data sets. The company's MuZero project, released in late 2020, improved on AlphaZero and learned how to perform a variety of tasks, including defeating AlphaZero at board games and mastering simple video games, all without ever actually being told the rules of any task or given any human-labeled training data.[28] Democracies should build on this and other technical foundations and lead the way in developing AI algorithms that reduce the need for data, negating one of the theorized autocratic advantages.[29]

In addition, some newer algorithms can retain the importance of training data but nonetheless preserve civil liberties and privacy. These machine learning techniques comprise only a tiny fraction of current AI research but show enormous potential.[30] One example is federated learning, in which an algorithm learns from data derived from many devices—such as users' cell phones—but does so in such a way that the data never leaves the user's control. Another fruitful area of research is differential privacy, in which algorithms introduce statistical noise to protect the privacy of users in a data set. Homomorphic encryption also offers promise, permitting disparate users to train machine learning algorithms on data without revealing underlying private information.

Although there are exceptions, the private sector mostly does not have incentives to pursue the fundamental research that can quickly develop and scale these new kinds of algorithms. Privacy-preserving machine learning is more nascent than traditional machine learning development. It also flies in the face of some advertising-centric business models. Democratic governments should step in, just as they have in other areas of technology, to catalyze research and innovation. The more compatible that AI can be made with decentralized technological and organizational structures, the more it will benefit democracies—and the weaker the claims that technology favors tyranny will become.

Likewise, democracies have to do more to ensure that computing power does not pose a barrier to open and innovative markets. As chapter 3 showed, large numbers of cutting-edge chips are often expensive, constraining advanced research. Even some of the most notable and best-funded AI labs in the world routinely max out their available computing resources. Getting access to sufficient power to train large and complex neural networks is a roadblock for many academic researchers and start-ups, and it is no exaggeration to say that there are likely many promising ideas in AI that have not been explored as a result.

In large part due to computing power's centrality, the high barriers to entry in the AI ecosystem are the mirror image of the low barriers to entry in the last technology boom. In the internet age, which powered decades of economic growth for democracies, it was easy for startups to get off the ground, quickly bringing their innovative ideas to market and toppling entrenched incumbents. Google, for example, was founded later than other major search engines, including Yahoo!, Lycos, Infoseek, AltaVista, and Ask Jeeves, and yet supplanted all of them with better technology. It will be harder for a future startup to replicate this leapfrogging feat in the age of AI. Such a static ecosystem works against the ideal functioning of dynamic markets and undermines constructive competition.

Democratic governments should make computing power more accessible to people with big ideas. Governments could buy cloud computing credits from leading companies at a large scale and at low rates. They could then grant access to those computing resources for academic labs or startups with broad social benefits, much as the National Science Foundation and Small Business Administration offer monetary grants to researchers and entrepreneurs. In addition to increasing innovation, making computational resources more accessible is essential for the age of AI, especially since many of the privacy-preserving or data-independent algorithms described above require substantial computational power to work.

More generally, democracies should acknowledge that the evangelists are right when they speak of the rapid pace of AI growth. The processes of government can sometimes be too slow to keep up with changing technology. For example, Jack Clark—then the policy director at OpenAI—noted that after GPT-2 came out, he arranged to brief government officials about its implications. By the time the meeting was eventually scheduled, OpenAI had invented the massively more powerful GPT-3.[31] Democracies need greater urgency in their AI strategy, and need to more proactively anticipate where the technology is going and how to get ahead of it. The status quo here is not good enough.

Democratic governments should also listen to the Cassandras. They are right to warn that machine learning is far more brittle than many proponents believe. Systems can easily fail in ways that are both unforeseen and dangerous. As machine learning systems increase in complexity and importance, concerns about reliability, robustness, and specification gaming will only grow. Democracies will overcome these concerns only with a substantial increase in research. One estimate holds that only 1 percent of machine learning research funding is dedicated to improving the systems' safety and security.[32] Once again, the private sector has shown that it does not have the incentive to lead the way, and democratic governments need to step up.

Other failures of machine learning systems should also give democratic policymakers pause. Bias tops the list. In many organizations, the most readily available training data set for any task is the record of previous human decisions. As Amazon learned with its résumé screening tool, and many other technology companies have learned with their products, these past decisions often exhibit tremendous bias, even if it is implicit. Machine learning systems, lacking any common sense or understanding, will learn those biases just as they learn any other pattern. Once deployed, they will cloak these biases in a false veneer of computational impartiality.

Similarly, the opacity of machine learning systems remains a major concern. Despite a great deal of technical research, these systems are in some respects getting more opaque, not less, as neural networks are employed to design neural networks.[33] As machine learning systems increase in complexity and scope, the reasons for their decisions will likely become even more inscrutable. Arbitrary decision-making might be par for the course in an autocracy, but the necessity of legitimate governance demands stricter scrutiny in a democracy. The need for oversight and transparency is especially urgent if governments or organizations that are central to modern life—such as banks, real estate agencies, insurance companies, and college admissions boards—use machine learning to aid their work.

On issues of bias, fairness, and transparency, the status quo is once again simply not good enough. Democracies need to be acutely aware of machine learning's limitations and must adjust their policies accordingly. The Trump administration's proposal that companies cannot be held liable for housing discrimination if they use an algorithm in their decisions offered an example of exactly the wrong approach, as the biases of such algorithms are quite insidious in their effects.[34] In the near future, a wide variety of government agencies and businesses will deploy machine learning across their organizations. Such a rollout has enormous potential to offer more efficient services and improved decision-making, but it must be done with proper attention to the limitations of AI systems.

Even if rendering a full explanation of decisions and proof of impartiality is not possible for neural networks, improved technology and transparency can build a stronger foundation of trust. For example, some organizations have taken to publishing information on the training data they provided to machine learning systems, offering more insight into the kinds of patterns that the algorithm learned.[35] Beyond that, governments should take the lead on investigating bias in machine learning systems and its effects; the US National Institute of Standards and Technology's audits of facial recognition systems offer a good example of the sort of work that democracies should pursue.[36]

More generally, for as much as democracies must worry about what AI will do in autocracies, they also must worry about what the technology will do at home. Even, or perhaps especially, when AI does not fail, it poses deep challenges to democracies. For example, powerful facial recognition technology and networked government surveillance systems should prompt renewed debates about the appropriate balance of privacy and security, as should such systems in the hands of citizens and corporations, which are no less powerful but largely unaccountable to democratic processes.[37] Even beyond facial recognition, democracies should generally consider how individuals' data is used, often without their knowledge or meaningful consent, to train machine learning systems. When it comes to disinformation, major political figures and widely followed social media accounts are not threats from autocracies, but nonetheless they can drive the same wedges and exploit the same digital terrain discussed in chapter 8. In each of these areas and many others, democracies have to choose where to adapt and where to remain steadfast against the new fire's force.

At home and abroad, democracies have an opportunity to shape the three sparks of the new fire—data, algorithms, and compute—so that they strengthen core values. The preceding chapters show how AI will shape statecraft and how statecraft will shape AI. Democracies should not sit back as this dynamic plays itself out. With proactive technological investments and values-driven policymaking, the United States and its allies can determine both what kind of AI develops and how people use it within and beyond their borders. All of this is necessary to prepare democracies to meet the autocratic challenge in the geopolitical arena.

GEOPOLITICS

In the age of AI, no democracy should be alone. Democracies' efforts are strongest when they coordinate with their allies.[38] The United States should lead a broad network of like-minded nations to share data, coordinate investments in privacy-enhancing technologies, enhance interoperability, and collaborate on AI research and development.[39] Governments should consider how best to deploy research funds in areas that will enable them to develop economically, innovate collaboratively, and strengthen democratic values. Formulating shared norms and standards for AI, such as the Principles on AI of the Organisation for Economic Cooperation and Development, is an excellent start, as are the efforts of the Global Partnership on AI. Democracies need to do the work required to translate these and other principles into a more concrete form that earns the trust of evangelists who are skeptical that governments are relevant and responsible in the age of AI.[40]

Democracies must collectively do better with setting international standards for AI than they have in other areas of technological advancement. China regularly advances its principles and objectives through multilateral forums, perhaps most notably on 5G standards. Russia, too, has sought to wield this kind of influence; for example, it has shaped the principles of the Shanghai Cooperation Organization on cybersecurity and has lobbied for these regional norms to become global. As standards continue to develop for key subjects in AI, such as facial recognition, reliability, and lethal autonomous weapons systems, democracies must work to lead the way.[41] They are off to a comparatively slow start, as China's government and companies have made an aggressive push to develop United Nations standards for facial recognition and surveillance systems.[42] Failure to lead on standards-setting processes cedes vital ground to autocracies that will be hard to regain.[43]

However fierce their competition with one another for talent, democracies must work together. If AI companies are more able to hire and easily move employees between allied democratic nations, all participants are likely to benefit. In addition, the United States should cultivate international networks of researchers through the exchange of knowledge and best practices. The National Science Foundation, for example, has awarded grants to researchers who promote international collaboration and benefit from the expertise and specialized skill sets of overseas partners; this investment is a foundation for future programs to strengthen ties between democracies. Longer term, democracies could lay the foundation for AI economic zones that enable researchers to work, study, and travel more freely between allied countries, provided they abide by a set of common principles on research security and technology protection.[44]

In a nod to the warriors, these partnerships should extend to the military. Just as allied governments have worked together on projects, such as developing new fighter aircraft, so too should they collaborate on using AI in their military operations. In 2020, the United States and twelve other nations established a foundation for such efforts by forming the AI Partnership for Defense. Facilitated by the Pentagon's Joint AI Center, this group of democracies established dialogues on both ethical matters, such as overarching principles, and practical ones, such as improving the interoperability and capability of AI systems.[45] These governments should continue to build on this foundation to strengthen their military and technological ties.

Hard policy choices are on the horizon. As some warriors have noted, it is challenging to predict the ethical, military, and strategic concerns that will flow from the development of lethal autonomous weapons. Democracies will have to continually reassess their posture as the technology evolves, but it is unwise to sit out the further responsible development of autonomy on the battlefield; it seems inevitable that AI's role in war will only increase. Intelligence gathering efforts should focus on how autocratic nations are using AI in peace and war so democracies are not caught flat-footed.

That said, the Cassandras are right to worry. Military commanders should have a deep appreciation for the limits of machine learning. They should know that AI is not nearly as capable or predictable in complex real-world environments as in controlled test settings—and few arenas are as complex as modern warfare. These deficiencies are not just confined to the areas of bias, opacity, and specification gaming but also include machine learning's susceptibility to hacking. Since the military applications of AI favored by the warriors are distinct from the scientific applications preferred by the evangelists, most machine learning systems are not ready for adversarial environments like war against opponents with good hacking capabilities; much more iterative testing and evaluation must be done before these systems can be considered ethically and strategically appropriate for combat.[46] Due to these major concerns about security and robustness, some high-stakes uses of machine learning—such as in nuclear command-and-control systems—are manifestly unwise for the foreseeable future.

Concerns around reliability and unintended escalation may offer some opportunity to work with authoritarian governments in a mutually beneficial way. Here, even though the technology differs, the arms control agreements and bilateral exchanges of the Cold War offer useful precedent and inspiration. In addition to the examples discussed in chapter 9, the Pugwash Conferences on Science and World Affairs fostered mutual understanding between US and Soviet scientists and engineers, beginning in the 1950s, while 1975's *Apollo-Soyuz* project brought together American and Soviet scientists for a joint mission in space.

Similar measures could lower the risks of miscalculation and inadvertent accidents today. Democracies should support dialogues on AI safety and security with Chinese and Russian counterparts at the governmental and nongovernmental levels. Scientists from these countries could also pursue low-stakes projects to build trust and transparency, such as understanding and translating each other's literature on AI safety and developing shared conceptions of how each side defines core ideas in the field. Democratic leaders should articulate—and live by—clear principles about their use of AI systems in peace and war, and they should offer public evidence of their commitments to robust testing and evaluation of these systems. Governments should create working groups of engineers who share testing protocols and should convene tabletop exercises among relevant officials to simulate the risks of machine learning systems in different contexts.[47]

For these confidence-building measures to work, democracies will need to overcome several hurdles evident in Chinese strategic thinking. Chinese leaders are loath to cast their country in the model of a withering Soviet Union; far from considering their country in relative decline, Chinese strategists perceive the trend lines as favoring Beijing, and therefore see no reason to constrain the growth of their military capabilities.[48] In response, democracies will need to strike a delicate balance between competition and cooperation, underscoring the dangers of miscalculation and misperception. Integrating AI into existing dialogues on strategic stability is one place to start.

Lastly, democracies must be ready to compete with autocracies in peacetime. Democracies have advantages too often overlooked by those who argue that technology favors tyranny. Besides the ability to attract talent, chief among these is that democracies lead the semiconductor industry. The continued Chinese dependence on democracies for computer chips and chip manufacturing equipment offers a substantial opportunity for policymakers. If democracies work together to limit Chinese access to chip manufacturing equipment, they can slow China's efforts to domestically develop advanced fabrication facilities and ensure the country remains dependent on imports for chips. Democracies can therefore maintain leverage over China's ability to build the advanced machine learning systems that are central to its perceived security and economic competitiveness as well as its expansive surveillance programs.[49]

The Trump White House threatened to ban the Chinese telecommunications giant ZTE from the US marketplace and choke off its supply of chips, backing down only when President Xi personally intervened.

In May 2019, the Department of Commerce placed restrictions on China's telecom giant Huawei, preventing the company from accessing any chips made anywhere in the world with US technology.[50] To bolster its domestic semiconductor manufacturing capacity, the United States pressured the Taiwanese chip maker TSMC to open fabrication facilities in the United States, pulling the company further into the American orbit. Congress allocated billions of dollars of incentives for the construction of fabrication facilities at home, part of a broader campaign to move the technology supply chain out of China and keep chip fabrication in democracies. Some policymakers in the United States debate going still further by placing more restrictive export controls on chips, attempting to turn a technical bottleneck into a chokehold.[51]

Controls on chips would be risky. There is a possibility that strong export controls might provide an advantage in the short term but lead to downsides in the long term. Cutting off China's access to chips would create an immediate incentive for China to invest even more in building a domestic industry. Given that Chinese buyers would in such circumstances have no access to foreign chips, the industry that China has been trying (and largely failing) to build just might finally flourish. American and allied chip companies would also see massive hits to their revenues, potentially constraining them from investing in the R&D that maintains their technological edge.

For this reason, it makes more sense to restrict China's imports of advanced chip manufacturing equipment but not restrict all Chinese companies from buying the chips themselves.[52] This strategy, which is the current US approach, lets China import chips from TSMC and other firms in democracies, though it tries to ban the use of chips for some purposes, such as surveillance—an area where more enforcement and greater allied cooperation on foreign science and technology monitoring would be valuable.[53] It does not permit China to import the advanced equipment used to manufacture competing products. The money from Chinese import of chips funds democracies' research into new and more advanced designs without bolstering the Chinese domestic chip industry.

It is not all rosy for democracies. While they should recognize their strengths and opportunities in export controls, they should also acknowledge their comparative vulnerability to disinformation. Even before the age of AI, autocracies employed disinformation to undermine the cred-ibility of democracies and fracture democratic alliances. The possibilities of deepfakes, automated propaganda efforts, and algorithmically boosted viral messages all amplify the threat of disinformation. There are no easy solutions here, especially given the extremely strong threat of homegrown disinformation. Efforts to stay ahead of the problem technologically are a good starting point, as are attempts to make algorithmic adjustments to limit the spread of disinformation. Ultimately, the best strategies will involve a mix of technical fixes, institutional reforms, educational invest-ments, and local and independent journalism. On this issue, as with so much else, the governments, private sectors, and civil societies of democ-racies will have to act quickly, decisively, and cooperatively.[54] It will not be easy.

THE NEW FIRE

On balance, democracies have a better hand to play than Harari and others suggest.[55] Their natural capacity to recruit and educate talent will make every part of their AI strategy more effective. They have strong advantages in technology, with realistic options available to maximize the potential of data, algorithms, and computing power across their societies. In so doing, democracies can shape the technological trajec-tory that AI will follow, steering its growth down a path that strengthens, rather than weakens, democratic values.

The geopolitical picture is more complex but still promising. Democ-racies enjoy clear strategic assets that they can wield in partnership with one another, such as powerful alliances and leverage over the semicon-ductor supply chain. In military applications of AI, ethical principles and reliability concerns will at times dictate that democracies be less aggres-sive than autocracies. But such trade-offs are not new. In intelligence, for example, democracies partly constrain their collection efforts to preserve some aspects of privacy; in warfare, democracies agree to limit their con-duct to accord with international law. Policymakers will have to navigate hard choices, but, with the right effort and perspective, they can do so.

What democracies—especially the United States—do not yet have is an appropriate sense of urgency, strategy, and follow-through. Efforts are nascent at best in nearly every area outlined above. The United States has almost entirely coasted on its previous reputation and investments for attracting talent, even as the rest of the world has worked assiduously to catch up. It has made only modest government investments in AI, missing opportunities at home to strengthen AI research and broaden access while also squandering opportunities abroad to set global norms and standards. Even in areas of comparative strength—such as the Department of Defense's creation of the Joint AI Center—much more work remains to be done on improving robustness, reliability, and standards.

The lack of urgency is especially damaging because of AI's prospects for growth in capability. Technological metrics show the increasing power of the combination of data, algorithms, and computing power. Many governments, not least that of the United States, have proven themselves ill-equipped to adapt to the pace of the age of AI. These governments are slow to address the crosscutting problems of the moment, slow to recognize that the nexus of invention has shifted in key respects to the private sector, and slow to regain the initiative. If the exponential growth in data, algorithms, and computing power continues, the costs of waiting to implement a meaningful strategy will only increase with each passing day and year.

The evangelists, warriors, and Cassandras offer their own visions for how to direct the new fire. No vision is fully correct or realistic, nor fully wrong. The evangelists deserve admiration for their devotion to progress toward creating a better world, especially with breakthroughs like Alpha-Fold, but they must recognize that most of their inventions will inevitably have geopolitical implications. These technological pioneers should take action to understand as best they can the consequences of what they invent, and to ensure that their creations benefit, rather than detract from, democratic prospects. On some issues, they will simply have to choose whether to side with democracy or autocracy.

The warriors are right to fear the threat from autocracies and smart to imagine how democracies could harness AI to defend against it. They should nonetheless try to earn the respect of the evangelists, whose technological talents are needed to better understand the AI tools at their disposal. They should also heed the warnings of the Cassandras, who offer necessary correctives about what AI can't do and some of the risks that follow. Though it may seem a distraction to some warriors, continual and mutually respectful engagement about ethical principles and technological limitations will only bring more strength and integrity to democracies' use of AI.

The Cassandras should recognize that the drumbeat of AI progress will continue no matter what they do. Indeed, as their name suggests, some of their warnings will never be heeded, even if they are right. Their mission should not be to thwart more capable AI entirely, but to temper evangelists' enthusiasm and challenge warriors' perspectives. They should seek both to understand the technology and to shape it, offering solutions to minimize its weaknesses even as its development races ahead.

All three groups should recognize that they are essential for democracy in the age of AI. The new fire has ignited, the fuel is plentiful, and the range of potential outcomes is vast. Some seek to harness the new fire, lighting a new course for civilization. Others seek to wield it, readying the flames for a war they hope never to have to fight. Still others seek to contain it, warning the world about the prospects of uncontrollable conflagration. And some—too many—ignore it, too preoccupied by other concerns to see the opportunities and dangers in front of them.

The new fire is striking in its complexity and its potential. AI raises questions that are more fundamental and more philosophical than the usual matters that dominate geopolitics. It forces a reexamination of big concepts, such as intelligence and understanding. It reshapes, for better or for worse, much of what it touches. But to focus only on the fire and its dancing flames is to miss the people in its shadows. These people are the ones who must decide how to tend it. How much light, warmth, and destruction the new fire brings depends on them. It depends on us.

ACKNOWLEDGMENTS

This is a book about choices, and perhaps the most important choice we made was to work on this project together. By happy coincidence, we had the idea for this book at the same time and pitched each other on the concept. Writing this book in partnership was one of the most enjoyable parts of this journey, even if COVID-19 meant that we almost never saw each other in person. We are grateful that our friendship not only survived the creative process but thrived as we pushed each other to think more deeply about the implications of artificial intelligence at a moment of uncertainty and profound transformations in the world.

This book would not have been possible without the support of our home institution: the School of Foreign Service, and in particular the Center for Security and Emerging Technology, at Georgetown University. Now in its second century, the SFS has a storied history, filled with faculty, staff, and students passionate about not just understanding the world but improving it. We are particularly grateful to SFS pillars Elizabeth Arsenault (and Crouton!), Dan Byman, Fr. Matthew Carnes, Christine Fair, Joel Hellman, Bruce Hoffman, Keir Lieber, Rebecca Patterson, and Elizabeth Stanley. Since its founding in January 2019, CSET has emerged as one of the most innovative centers of research on issues at the intersection of national security and emerging technologies. Its success is due in no small measure to the generosity and strategic vision of its founding director, Jason Matheny. Anyone who knows Jason will agree that he sets the standard for brilliant insight and deep kindness, and we are enormously fortunate to benefit from his encyclopedic knowledge, thoughtful mentorship, and policy expertise. This book would not exist without him.

Many individuals helped us to sharpen and support our arguments, but none more so than Beba Cibralic and Ryan Fedasiuk, two fast-rising academic and policy stars. Beba is a marvel. She brought a philosopher's keen sense of structure and rigor to this project, talking arguments through with us, pushing us to consider ethical questions more deeply, and suggesting fruitful lines of research. She offered a razor-sharp eye for language, reading nearly every draft chapter and helping us tighten its prose. In addition, she demonstrated a remarkable ability to find telling details, enlivening the people in this book and doing justice to their views on complex issues. This book simply would not be what it is without her, and we cannot wait to see what she does next.

Ryan Fedasiuk is what baseball scouts would call a "five-tool player," capable of far-sighted strategic thinking, in-depth case study work, data analysis, research support, and rapid-fire fact-checking. Somehow Ryan managed to support us on this project while juggling graduate school and a full-time job at CSET. His ability to delve into new literature and distill the most essential points seemingly before we hit the "send" button on our emails was as astounding as it was critical for this book. His intellectual curiosity, love of languages, and careful study of primary sources is evident in the impressive body of research he has already published at CSET and elsewhere. Above all, we valued Ryan's sharp eye and willingness to challenge us on every argument, concept, and recommendation in these pages. There is no doubt that this book is stronger for Ryan's natural aptitude for red teaming our fundamental assumptions, and we know great things are ahead for him.

In addition to Beba and Ryan, we relied on other minds far wiser than ours in our effort to encapsulate some of the major developments in AI over the past several decades. As one look at the citations will attest, we are indebted to all of our CSET colleagues. This includes the many individuals on whose research we have drawn and who gave generously of their time to review all or parts of the manuscript or otherwise aid in its production. Special thanks to Olivia Albrighton-Vanway, Tessa Baker, Jack Clark, Saif Khan, Margarita Konaev, Drew Lohn, Igor Mikolic-Torreira, Dewey Murdick, Chris Rohlf, Helen Toner, and the entire CyberAI team. Teddy Collins and John Bansemer both offered extremely detailed readings of the book, while Tarun Chhabra went above and beyond with his insights and kindness. Outside of CSET, we are grateful to Perri Adams, Lisa Wiswell Coe, Fiona Cunningham, Jeffrey Ding, Rebekah Kennel, Luke Muehlhauser, and Alex Palmer for their insights, as well as to all of our interview subjects. As ever, all errors remain ours alone and the opinions and characterizations in this book do not necessarily represent those of the US government.

The Massachusetts Institute of Technology has long been at the forefront of some of the most consequential breakthroughs in AI, and the MIT Press has provided a voice and venue to many of the leading lights in the field. Our deepest thanks to Gita Devi Manaktala, Judy Feldmann, and their entire team for bringing this project to fruition. The decision to work with Gita was one of the easiest for us to make. She grasped effortlessly what this book is about and why AI matters for understanding the world today and the world to come. We are grateful to Yasuyo Iguchi and Mary Reilly for helping with the design and graphics of the book and to Erika Barrios and Pamela Quick for copyright assistance. We also wish to thank the three anonymous reviewers of our manuscript for their excellent suggestions. In addition, Brian Bergstein, Rachel Fudge, Rebecca Kessler, and Julia Ringo provided superb editorial insights, helping us refine the text, while Matt Mahoney and Dale Brauner brought their customary eagle eyes to every fact and assertion.

Our agents believed in this book from the start. For Ben, Michael Carlisle and William Callahan of Inkwell Management have offered shrewd advice for years. For Andrew, Bridget Matzie has been a gem. She has a knack for seeing the bigger picture and for always pushing her authors to communicate their ideas with clarity and through memorable characters. We are grateful that our respective agents worked together so seamlessly.

Ben deeply appreciates all the people who have helped him refine his thinking on technology and security over the years, and who have made life so wonderful. Thomas Rid is a great scholar and an even better friend. Michael Sulmeyer is a treasure to his country and to all those around him.

At the Wilson Center, Meg King is a get-things-done rockstar. Friends and teachers from Regis, Georgetown, and the Marshall Scholarship know how meaningful they are, especially as one decade of friendship turns into two, unencumbered by time and distance. Family deserves its own special recognition. Kelly is a perfect partner in this project and in life. As ever, her uproarious laugh brightens the day of all who hear it. Laura and Bruce are wonderful parents full of unconditional love, while Gerard and Annie are the perfect godparents. Other family members—including Fayez, Susan, Tracy, and Laurie—have all cheerfully listened to way too many conversations about AI. Last but not least is Mary, an excellent editor in her own right and the best younger sister one can ask for. With gratitude for her insight and love, Ben dedicates this book to her.

Andrew is grateful to his loving parents, Will and Maris, and his wonderfully creative brother, Chris. He cherishes the memory of two special people: Len Hawley and Evelyn Osterweil. Len was the most selfless mentor one could ask for in life. Even in his last weeks, he was always thinking of others, sharing an insight here or passing on a worthwhile article there. He lived a life of purpose and public service, investing himself so fully in everyone he met and every project in which he was engaged. Evelyn was likewise a source of constant inspiration and *joie de vivre* for Andrew and his family. Her memory will always be a blessing for the many people whom she touched with her elegance and artistry. Andrew's wife, Teresa, is an endless source of joy whose love of life, books, and music has been a shining light even in the darkest days of a pandemic. He dedicates this book to her.

NOTES

INTRODUCTION

1. Karina Fonseca-Azevedo and Suzana Herculano-Houzel, "Metabolic Constraint Imposes Tradeoff between Body Size and Number of Brain Neurons in Human Evolution," *Proceedings of the National Academy of Sciences of the United States of America* 109, no. 45 (November 6, 2012): 18571–18676.

2. John Pryor and Elizabeth M. Jeffreys, "The Age of the ΔΡΟΜΩΝ: The Byzantine Navy ca. 500–1204," in *The Age of the ΔΡΟΜΩΝ* (Boston: Brill, 2006), 607–609.

3. Catherine Jewell, "Artificial Intelligence: The New Electricity," *WIPO Magazine*, June 2019, https://www.wipo.int/wipo_magazine/en/2019/03/article_0001.html; Andrew Ng, "Artificial Intelligence Is the New Electricity," Stanford Graduate School of Business MSx Future Forum, January 25, 2017, YouTube video, 1:27:43, https://www.youtube.com/watch?v=21EiKfQYZXc.

4. Ernest Freeberg, *The Age of Edison: Electric Light and the Invention of Modern America* (London: Penguin, 2013), 9. It is worth noting that some, like Google's Sundar Pichai, use electricity and fire interchangeably as metaphors, though not particularly with reference to the broad range of outcomes fire offers.

5. Ray Kurzweil, *The Singularity Is Near: When Humans Transcend Biology* (New York: Viking Press, 2005). We believe that intelligence is best viewed not as crossing a single threshold, but rather as encompassing a spectrum of possibilities. In one respect, intelligence is a task-dependent measure on which comparison to humans does not always shed light; we should expect AI to excel at some tasks, such as Go, even as it struggles at others, such as charades. In another respect, intelligence is characterized by an ability to generalize between tasks with ease and skill and learn a broad range of new tasks with flexibility. Shane Legg and Marcus Hutter, "A Collection of Definitions of Intelligence," arXiv, June 15, 2007, https://arxiv.org/pdf

/0706.3639.pdf; François Chollet, "On the Measure of Intelligence," arXiv, November 5, 2019, https://arxiv.org/pdf/1911.01547.pdf.

6. David Edgerton, *The Shock of the Old: Technology and Global History since 1900* (Oxford: Oxford University Press, 2011); William H. McNeill, *The Pursuit of Power: Technology, Armed Force, and Society since A.D. 1000* (Chicago: University of Chicago Press, 2013).

7. Ben Buchanan, *The AI Triad and What It Means for National Security Strategy* (Georgetown University Center for Security and Emerging Technology, August 2020), https://cset.georgetown.edu/wp-content/uploads/CSET-AI-Triad-Report.pdf; Dewey Murdick, James Dunham, and Jennifer Melot, *AI Definitions Affect Policymaking* (Georgetown University Center for Security and Emerging Technology, June 2, 2020), https://cset.georgetown.edu/wp-content/uploads/CSET-AI-Definitions-Affect -Policymaking.pdf.

8. Gregory C. Allen, *Understanding China's AI Strategy*, Center for a New American Security, February 6, 2019, https://www.cnas.org/publications/reports/understanding -chinas-ai-strategy; Elsa B. Kania, *"AI Weapons" in China's Military Innovation* (Brookings Institution, April 2020), https://www.brookings.edu/wp-content/uploads/2020/04 /FP_20200427_ai_weapons_kania_v2.pdf; Lingling Wei, "China's Xi Ramps Up Control of Private Sector. 'We Have No Choice but to Follow the Party,'" *Wall Street Journal*, December 10, 2020, https://wsj.com/articles/china-xi-clampdown-private-sector -communist-party-11607612531.

9. Melissa K. Chan, "Could China Develop Killer Robots in the Near Future? Experts Fear So," *Time*, September 13, 2019, https://time.com/5673240/china-killer -robots-weapons/.

10. Yuval Noah Harari, "Why Technology Favors Tyranny," *Atlantic*, August 30, 2018, https://www.theatlantic.com/magazine/archive/2018/10/yuval-noah-harari -technology-tyranny/568330/.

11. Andrew Imbrie and Elsa B. Kania, *AI Safety, Security, and Stability among the Great Powers* (Georgetown University Center for Security and Emerging Technology, December 2019), https://cset.georgetown.edu/wp-content/uploads/AI-Safety -Security-and-Stability-Among-the-Great-Powers.pdf.

CHAPTER 1

1. "AFI's 100 Years . . . 100 Heroes and Villains," American Film Institute, accessed July 1, 2020, https://www.afi.com/afis-100-years-100-heroes-villians/.

2. Technically, these systems manipulate representations of words, not words themselves.

3. Michele Banko and Eric Brill, "Scaling to Very Very Large Corpora for Natural Language Disambiguation," *Proceedings of the 39th Annual Meeting on Association for Computational Linguistics* (Stroudsburg, PA: Association for Computational Linguistics, 2001), 26–33.

4. Chris Anderson et al., "The End of Theory: The Data Deluge Makes the Scientific Method Obsolete," *Wired*, June 23, 2008, https://www.wired.com/2008/06/pb -theory/.

5. Scott Cleland, "Google's 'Infringenovation' Secrets," *Forbes*, October 3, 2011, https://www.forbes.com/sites/scottcleland/2011/10/03/googles-infringenovation -secrets/#7bff9df030a6.

6. Maria Deutscher, "IBM's CEO Says Big Data Is Like Oil, Enterprises Need Help Extracting the Value," SiliconANGLE, March 11, 2013, https://siliconangle.com /2013/03/11/ibms-ceo-says-big-data-is-like-oil-enterprises-need-help-extracting-the -value/.

7. Kai-Fu Lee, *AI Superpowers: China, Silicon Valley, and the New World Order* (Boston: Houghton Mifflin Harcourt, 2018), 55.

8. Caleb Garling, "Andrew Ng: Why 'Deep Learning' Is a Mandate for Humans, Not Just Machines," *Wired*, May 5, 2015, https://www.wired.com/brandlab/2015/05 /andrew-ng-deep-learning-mandate-humans-not-just-machines/.

9. Michael Palmer, "Data Is the New Oil," ANA Marketing Maestros, November 3, 2006, https://ana.blogs.com/maestros/2006/11/data_is_the_new.html; Charles Arthur, "Tech Giants May Be Huge, but Nothing Matches Big Data," *Guardian*, August 23, 2013, http://www.theguardian.com/technology/2013/aug/23/tech-giants-data.

10. Chris Dixon, "How Aristotle Created the Computer," *Atlantic*, March 20, 2017, https://www.theatlantic.com/technology/archive/2017/03/aristotle-computer /518697/.

11. The term itself came from John McCarthy, one of the participants in the Dartmouth project.

12. Dorothy Leonard-Barton and John Sviokla, "Putting Expert Systems to Work," *Harvard Business Review*, March 1, 1988, https://hbr.org/1988/03/putting -expert-systems-to-work.

13. In the case of color images, several neurons often correspond to a single pixel.

14. Often, an additional mathematical operation—known as an activation function—is used here as well to determine the value of the next layer's neuron.

15. There is an additional number called the "bias term" that is not necessary for this conceptual explanation. For more on neural networks, see Vishal Maini and Samer Sabri, "Machine Learning for Humans" (self-published report, September 2017), https://www.dropbox.com/s/e38nil1dnl7481q/machine_learning.pdf?dl=0.

16. Humans do often set the value of variables known as hyperparameters, which affect how the network learns, but discussion of hyperparameters is not necessary for this conceptual explanation.

17. It is worth noting that Leibniz also invented calculus, which is used in this process of training neural networks.

18. Jessi Hempel, "Fei-Fei Li's Quest to Make AI Better for Humanity," *Wired*, November 13, 2018, https://www.wired.com/story/fei-fei-li-artificial-intelligence-humanity/.

19. Francesca Simion and Elisa Di Giorgio, "Face Perception and Processing in Early Infancy: Inborn Predispositions and Developmental Changes," *Frontiers in Psychology* 6 (July 9, 2015): 969.

20. Hempel, "Fei-Fei Li's Quest to Make AI Better for Humanity."

21. Adit Deshpande, "A Beginner's Guide to Understanding Convolutional Neural Networks," GitHub, July 20, 2016, https://adeshpande3.github.io/A-Beginner%27s -Guide-To-Understanding-Convolutional-Neural-Networks/.

22. As noted above, AlexNet, like most neural networks, includes both weights and biases as parameters, but we simplify the discussion here for readability.

23. Alex Krizhevsky, Ilya Sutskever, and Geoffrey E. Hinton, "ImageNet Classification with Deep Convolutional Neural Networks," in *Proceedings of the 25th International Conference on Neural Information Processing Systems* (Red Hook, NY: Curran Associates, 2012), 1097–1105.

24. Luciano Strika, "Feature Visualization on Convolutional Neural Networks (Keras)," Towards Data Science, May 30, 2020, https://towardsdatascience.com /feature-visualization-on-convolutional-neural-networks-keras-5561a116d1af; Krizhevsky, Sutskever, and Hinton, "ImageNet Classification with Deep Convolutional Neural Networks."

25. Aaron Tilley, "China's Rise in the Global AI Race Emerges as It Takes Over the Final ImageNet Competition," *Forbes*, July 31, 2017, https://www.forbes.com/sites /aarontilley/2017/07/31/china-ai-imagenet/.

26. Robert McMillan, "This Guy Beat Google's Super-Smart AI—But It Wasn't Easy," *Wired*, January 21, 2015, https://www.wired.com/2015/01/karpathy/.

27. Matt Sheehan, "Who Benefits From American AI Research in China?," Macro-Polo, October 21, 2019, https://macropolo.org/china-ai-research-resnet/; Kaiming He, LinkedIn profile, accessed August 12, 2020, https://www.linkedin.com/in/kaiming -he-90664838.

28. "ImageNet Large Scale Visual Recognition Competition (ILSVRC)," ImageNet, 2015, http://www.image-net.org/challenges/LSVRC/; Jason Brownlee, "A Gentle Introduction to the ImageNet Challenge (ILSVRC)," Machine Learning Mastery, April 30, 2019, https://machinelearningmastery.com/introduction-to-the-imagenet -large-scale-visual-recognition-challenge-ilsvrc/; Arthur Ouaknine, "Review of Deep Learning Algorithms for Image Classification," Zyl Story, January 17, 2018, https:// medium.com/zylapp/review-of-deep-learning-algorithms-for-image-classification -5fdbca4a05e2; Tilley, "China's Rise in the Global AI Race Emerges."

29. Michelle Yeo, Tristan Fletcher, and John Shawe-Taylor, "Machine Learning in Fine Wine Price Prediction," *Journal of Wine Economics* 10, no. 2 (2015): 151–172.

30. "Session with Ian Goodfellow," Quora, August 12, 2016, https://www.quora .com/session/Ian-Goodfellow/1.

31. Martin Giles, "The GANfather: The Man Who's Given Machines the Gift of Imagination," *MIT Technology Review*, February 21, 2018, https://www.technology

review.com/2018/02/21/145289/the-ganfather-the-man-whos-given-machines-the
-gift-of-imagination/.

32. "Session with Ian Goodfellow."

33. "Session with Ian Goodfellow."

34. "Session with Ian Goodfellow."

35. "Session with Ian Goodfellow."

36. Veronika Cheplygina, "How I Fail S01E21: Ian Goodfellow," May 7, 2018, https://
veronikach.com/how-i-fail/how-i-fail-ian-goodfellow-phd14-computer-science/.

37. Allison Toh, "An Argument in a Bar Led to the Generative Adversarial Networks,"
Nvidia, June 8, 2017, https://blogs.nvidia.com/blog/2017/06/08/ai-podcast-an
-argument-in-a-bar-led-to-the-generative-adversarial-networks-revolutionizing-deep
-learning/.

38. Jamie Beckett, "What's a Generative Adversarial Network? Inventor Explains,"
Nvidia, May 17, 2017, https://blogs.nvidia.com/blog/2017/05/17/generative-adversarial
-networks/.

39. Giles, "The GANfather."

40. Beckett, "What's a Generative Adversarial Network?"

41. GANs are more often used for photos and videos than music, but they can
create all three.

42. Ian J. Goodfellow et al., "Generative Adversarial Networks," arXiv, June 10,
2014, http://arxiv.org/abs/1406.2661; "Generative Adversarial Networks: Overview
of GAN Structure," Google Developers, accessed July 1, 2020, https://developers
.google.com/machine-learning/gan/gan_structure; Bernard Marr, "Artificial Intelligence Explained: What Are Generative Adversarial Networks (GANs)?," *Forbes*, June
12, 2019, https://www.forbes.com/sites/bernardmarr/2019/06/12/artificial-intelligence
-explained-what-are-generative-adversarial-networks-gans/.

43. Ian Goodfellow, "4.5 years of GAN Progress on Face Generation," Twitter,
January 15, 2019, 4:40 p.m., https://twitter.com/goodfellow_ian/status/108497359
6236144640.

44. Rob Toews, "Deepfakes Are Going to Wreak Havoc on Society: We Are Not Prepared," *Forbes*, May 25, 2020, https://www.forbes.com/sites/robtoews/2020/05/25
/deepfakes-are-going-to-wreak-havoc-on-society-we-are-not-prepared/; Ben Beaumont-
Thomas, "Jay-Z Takes Action against 'Deepfakes' of Him Rapping Hamlet and Billy
Joel," *Guardian*, April 29, 2020, http://www.theguardian.com/music/2020/apr/29/jay
-z-files-takes-action-against-deepfakes-of-him-rapping-hamlet-and-billy-joel.

45. Raphael Satter, "Experts: Spy Used AI-Generated Face to Connect with Targets,"
AP News, June 14, 2019, https://apnews.com/bc2f19097a4c4fffaa00de6770b8a60d;
Benjamin Strick, "West Papua: New Online Influence Operation Attempts to Sway
Independence Debate," *Bellingcat*, November 11, 2020, https://www.bellingcat.com
/news/2020/11/11/west-papua-new-online-influence-operation-attempts-to-sway
-independence-debate/; *Fake Cluster Boosts Huawei* (Graphika, January 2021), https://

public-assets.graphika.com/reports/graphika_report_fake_cluster_boosts_huawei .pdf.

46. The GAN created the video, while the audio of the speech was performed by a human impersonator. Charlotte Jee, "An Indian Politician Is Using Deepfake Technology to Win New Voters," *MIT Technology Review*, February 19, 2020, https://www .technologyreview.com/2020/02/19/868173/an-indian-politician-is-using-deepfakes -to-try-and-win-voters/.

CHAPTER 2

1. This comparison does not include the Xia dynasty, a part of traditional Chinese historiography but one for which there are no contemporaneous records.

2. In essence, these new operations were new one-step algorithms.

3. L. F. Menabrea and Ada Augusta Lovelace, *Sketch of the Analytical Engine Invented by Charles Babbage, Esq*, vol. 3 (1842; repr., London: Scientific Memoirs, 2014).

4. Danny Hernandez and Tom B. Brown, "Measuring the Algorithmic Efficiency of Neural Networks," arXiv, May 8, 2020, http://arxiv.org/abs/2005.04305.

5. Robert Wright, "Can Machines Think?," *Time*, March 25, 1996, http://content .time.com/time/magazine/article/0,9171,984304,00.html.

6. Steven Levy, "What Deep Blue Tells Us about AI in 2017," *Wired*, May 23, 2017, https://www.wired.com/2017/05/what-deep-blue-tells-us-about-ai-in-2017/; Rajiv Chandrasekaran, "Kasparov Proves No Match for Computer," *Washington Post*, May 12, 1997, https://www.washingtonpost.com/wp-srv/tech/analysis/kasparov/kasparov .htm.

7. Dag Spicer, "Oral History of Feng-Hsiung Hsu," Computer History Museum, February 14, 2005, 9, http://archive.computerhistory.org/resources/text/Oral_History /Hsu_Feng_Hsiung/hsu.oral_history_transcript.2005.102657920.pdf.

8. BBC News, "Deep Blue vs Kasparov: How a Computer Beat Best Chess Player in the World," YouTube video, May 14, 2017, 2:00, https://www.youtube.com /watch?v=KF6sLCeBj0s.

9. Luke Harding and Leonard Barden, "From the Archive, 12 May 1997: Deep Blue Win a Giant Step for Computerkind," *Guardian*, May 12, 2011, http://www .theguardian.com/theguardian/2011/may/12/deep-blue-beats-kasparov-1997.

10. Mark Robert Anderson, "Twenty Years on from Deep Blue vs Kasparov: How a Chess Match Started the Big Data Revolution," *Conversation*, May 11, 2017, http:// theconversation.com/twenty-years-on-from-deep-blue-vs-kasparov-how-a-chess -match-started-the-big-data-revolution-76882.

11. Bruce Weber, "Swift and Slashing, Computer Topples Kasparov," *New York Times*, May 12, 1997, https://www.nytimes.com/1997/05/12/nyregion/swift-and-slashing -computer-topples-kasparov.html; Jack Peters, "After Sudden Defeat, It's Kasparov Who's Blue," *Los Angeles Times*, May 12, 1997, https://www.latimes.com/archives/la -xpm-1997-05-12-mn-57996-story.html.

12. Patrick House, "The Electronic Holy War," *New Yorker*, March 25, 2014, https://www.newyorker.com/tech/annals-of-technology/the-electronic-holy-war.

13. Greg Kohs, dir., *AlphaGo—The Movie* (Moxie Pictures, September 29, 2017), YouTube video, 1:30:27, https://www.youtube.com/watch?v=WXuK6gekU1Y.

14. John Tromp, "The Number of Legal Go Positions," in *Computers and Games*, ed. Aske Plaat, Walter Kosters, and Jaap van den Herik (Cham: Springer International, 2016), 183–190; J. Richard Gott III et al., "A Map of the Universe," *Astrophysical Journal* 624, no. 2 (2005): 463.

15. Gary A. Klein, Roberta Calderwood, and Anne Clinton-Cirocco, "Rapid Decision Making on the Fire Ground," *Proceedings of the Human Factors Society Annual Meeting* 30, no. 6 (1986), https://journals.sagepub.com/doi/abs/10.1177/154193128603000616; Gary A. Klein, "A Recognition-Primed Decision (RPD) Model of Rapid Decision Making," in *Decision Making in Action: Models and Methods*, ed. Gary Klein et al. (New York: Ablex, 1993), https://psycnet.apa.org/record/1993-97634-006.

16. Nature, "The Computer That Mastered Go," YouTube video, January 27, 2016, 3:05, https://www.youtube.com/watch?v=g-dKXOlsf98.

17. "Four Arts," China Online Museum, accessed July 1, 2020, https://comuseum.com/culture/four-arts/.

18. George Johnson, "To Test a Powerful Computer, Play an Ancient Game," *New York Times*, July 29, 1997, https://www.nytimes.com/1997/07/29/science/to-test-a-powerful-computer-play-an-ancient-game.html.

19. Clemency Burton-Hill, "The Superhero of Artificial Intelligence: Can This Genius Keep It in Check?," *Guardian*, February 16, 2016, http://www.theguardian.com/technology/2016/feb/16/demis-hassabis-artificial-intelligence-deepmind-alphago.

20. Hal Hodson, "DeepMind and Google: The Battle to Control Artificial Intelligence," *Economist*, March 1, 2019, https://www.economist.com/1843/2019/03/01/deepmind-and-google-the-battle-to-control-artificial-intelligence.

21. Burton-Hill, "The Superhero of Artificial Intelligence?"

22. Silver had begun his PhD the year prior, in 2004.

23. David Silver, "Reinforcement Learning and Simulation-Based Search in Computer Go" (PhD diss., University of Alberta, 2009).

24. It is worth noting that this description refers to deep reinforcement learning, and some simpler agents do not use neural networks. Leslie Pack Kaelbling, Michael L. Littman, and Andrew W. Moore, "Reinforcement Learning: A Survey," *Journal of Artificial Intelligence Research* 4 (1996): 237–285; Richard S. Sutton and Andrew G. Barto, *Reinforcement Learning: An Introduction*, 2nd ed. (Cambridge, MA: MIT Press, 2018); Volodymyr Mnih et al., "Human-Level Control through Deep Reinforcement Learning," *Nature* 518, no. 7540 (2015): 529–533.

25. Tom Simonite, "How Google Plans to Solve Artificial Intelligence," *MIT Technology Review*, March 31, 2016, https://www.technologyreview.com/2016/03/31/161234/how-google-plans-to-solve-artificial-intelligence/. For more on artificial general

intelligence, see Nick Bostrom, *Superintelligence* (Oxford: Oxford University Press, 2014).

26. Burton-Hill, "The Superhero of Artificial Intelligence."

27. DeepMind, "DeepMind: The Podcast | Episode 8: Demis Hassabis—The Interview," YouTube video, March 4, 2020, 05:00–06:40, https://www.youtube.com/watch?v=vcLU0DhDhi0.

28. DeepMind, "DeepMind: The Podcast," 10:46–15:24; Cade Metz, "What the AI Behind AlphaGo Can Teach Us about Being Human," *Wired*, May 17, 2016, https://www.wired.com/2016/05/google-alpha-go-ai/.

29. Mnih et al., "Human-Level Control through Deep Reinforcement Learning."

30. Jordan Novet, "Google Is Finding Ways to Make Money from Alphabet's Deep-Mind A.I. Technology," *CNBC*, March 31, 2018, https://www.cnbc.com/2018/03/31/how-google-makes-money-from-alphabets-deepmind-ai-research-group.html.

31. Geordie Wood, "In Two Moves, AlphaGo and Lee Sedol Redefined the Future," *Wired*, March 16, 2016, https://www.wired.com/2016/03/two-moves-alphago-lee-sedol-redefined-future/.

32. David Silver et al., "Mastering the Game of Go with Deep Neural Networks and Tree Search," *Nature* 529, no. 7587 (2016): 484–489; "AlphaGo," DeepMind, accessed July 1, 2020, https://deepmind.com/research/case-studies/alphago-the-story-so-far; Jonathan Hui, "AlphaGo: How It Works Technically?," Medium, May 13, 2018, https://medium.com/@jonathan_hui/alphago-how-it-works-technically-26ddcc085319.

33. Silver et al., "Mastering the Game of Go with Deep Neural Networks and Tree Search."

34. Kohs, *AlphaGo*, 14:36–14:48.

35. Kohs, *AlphaGo*, 31:16–31:18.

36. Kohs, *AlphaGo*, 27:52–27:58.

37. Kohs, *AlphaGo*, 34:50–34:57.

38. John D. Kelleher, *Deep Learning* (Cambridge, MA: MIT Press, 2019), 2–4.

39. Christopher Moyer, "How Google's AlphaGo Beat a Go World Champion," *Atlantic*, March 28, 2016, https://www.theatlantic.com/technology/archive/2016/03/the-invisible-opponent/475611/.

40. Kohs, *AlphaGo*, 52:43–52:54.

41. Cade Metz, "The Sadness and Beauty of Watching Google's AI Play Go," *Wired*, November 16, 2003, https://www.wired.com/2016/03/sadness-beauty-watching-googles-ai-play-go/.

42. Kohs, *AlphaGo*, 53:06–53:15.

43. Kohs, *AlphaGo*, 55:52–55:59.

44. Kohs, *AlphaGo*, 58:29–58:52.

45. Kohs, *AlphaGo*, 63:30–63:35.

46. Kohs, *AlphaGo*, 62:32–62:58.

47. Kohs, *AlphaGo*, 61:27–61:38.

48. Kohs, *AlphaGo*, 74:25–74:31.

49. Wood, "In Two Moves, AlphaGo and Lee Sedol Redefined the Future."

50. Kohs, *AlphaGo*, 81:01–81:05.

51. Kohs, *AlphaGo*, 83:13.

52. Demis Hassabis, "Excited to share an update on #AlphaGo!," Twitter, January 5, 2017, 7:00 a.m., https://twitter.com/demishassabis/status/816660463282954240.

53. "Humans Mourn Loss after Google Is Unmasked as China's Go Master," *Wall Street Journal*, January 5, 2017, https://www.wsj.com/articles/ai-program-vanquishes -human-players-of-go-in-china-1483601561.

54. Zheping Huang, "The AI 'Master' Bested the World's Top Go Players—and Then Revealed Itself as Google's AlphaGo in Disguise," *Quartz*, January 4, 2017, https:// qz.com/877721/the-ai-master-bested-the-worlds-top-go-players-and-then-revealed -itself-as-googles-alphago-in-disguise/; Lee, *AI Superpowers*, 5.

55. Sam Byford, "AlphaGo Retires from Competitive Go after Defeating World Number One 3–0," *Verge*, May 27, 2017, https://www.theverge.com/2017/5/27 /15704088/alphago-ke-jie-game-3-result-retires-future.

56. Silver et al., "Mastering the Game of Go with Deep Neural Networks and Tree Search"; DeepMind, "AlphaGo"; Hui, "AlphaGo: How It Works Technically?"; Kohs, *AlphaGo*.

57. Steven Strogatz, "One Giant Step for a Chess-Playing Machine," *New York Times*, December 26, 2018, https://www.nytimes.com/2018/12/26/science/chess-artificial -intelligence.html.

58. Silver et al., "Mastering the Game of Go with Deep Neural Networks and Tree Search"; DeepMind, "AlphaGo"; Hui, "AlphaGo: How It Works Technically?"; Kohs, *AlphaGo*.

59. "'Superhuman' Google AI Claims Chess Crown," *BBC*, December 6, 2017, https://www.bbc.com/news/technology-42251535.

60. David Silver et al., "AlphaZero: Shedding New Light on the Grand Games of Chess, Shogi and Go," DeepMind, December 6, 2018, https://deepmind.com /blog/article/alphazero-shedding-new-light-grand-games-chess-shogi-and-go; Nenad Tomašev et al., "Assessing Game Balance with AlphaZero: Exploring Alternative Rule Sets in Chess," arXiv, September 9, 2020, http://arxiv.org/abs/2009.04374.

61. Demis Hassabis, "The Power of Self-Learning Systems" (video), MIT Center for Brains, Minds, and Machines, March 20, 2019, https://cbmm.mit.edu/video/power -self-learning-systems.

62. David Silver et al., "A General Reinforcement Learning Algorithm That Masters Chess, Shogi, and Go through Self-Play," *Science* 362, no. 6419 (December 7, 2018): 1140–1144. It is worth noting that these training runs were three separate events.

63. In 2020, DeepMind released MuZero, a continuation of the AlphaZero research agenda. As discussed in chapter 10, the improved system can learn how to achieve many different tasks, including defeating AlphaZero at Go, without being told the rules of the task. Julian Schrittwieser et al., "Mastering Atari, Go, Chess and Shogi by Planning with a Learned Model," *Nature* 588, no. 7839 (December 2020): 604–609.

64. For an early articulation of China's Sputnik moment, see Lee, *AI Superpowers*, ch. 1.

65. John Noble Wilford, "With Fear and Wonder in Its Wake, Sputnik Lifted Us into the Future," *New York Times*, September 25, 2007, https://www.nytimes.com /2007/09/25/science/space/25sput.html.

66. Zachary Arnold and Ashwin Acharya, *Chinese Public AI R&D Spending: Provisional Findings* (Georgetown University Center for Security and Emerging Technology, December 2019), https://cset.georgetown.edu/wp-content/uploads/Chinese-Public -AI-RD-Spending-Provisional-Findings-2.pdf.

67. James Vincent, "China and the US Are Battling to Become the World's First AI Superpower," *Verge*, August 3, 2017, https://www.theverge.com/2017/8/3/16007736 /china-us-ai-artificial-intelligence; Matt Schiavenza, "China's 'Sputnik Moment' and the Sino-American Battle for AI Supremacy," Asia Society, September 25, 2018, https://asiasociety.org/blog/asia/chinas-sputnik-moment-and-sino-american-battle -ai-supremacy; Andrew Imbrie, Elsa B. Kania, and Lorand Laskai, *The Question of Comparative Advantage in Artificial Intelligence* (Georgetown University Center for Security and Emerging Technology, January 2020), https://cset.georgetown.edu /research/the-question-of-comparative-advantage-in-artificial-intelligence-enduring -strengths-and-emerging-challenges-for-the-united-states/; Allen, *Understanding China's AI Strategy*, 6.

68. Cade Metz, "Making New Drugs with a Dose of Artificial Intelligence," *New York Times*, February 5, 2019, https://www.nytimes.com/2019/02/05/technology /artificial-intelligence-drug-research-deepmind.html.

69. The word *protein* is from the Greek for "of primary importance."

70. Greg Williams, "Inside DeepMind's Epic Mission to Solve Science's Trickiest Problem," *Wired*, August 6, 2019, https://www.wired.co.uk/article/deepmind-protein -folding.

71. For a different version of this pearl analogy, see Alistair Martin, "Introduction to the Protein Folding Problem," Oxford Protein Informatics Group, May 2, 2014, https:// www.blopig.com/blog/2014/05/introduction-to-the-protein-folding-problem/.

72. Patrick Sweeney et al., "Protein Misfolding in Neurodegenerative Diseases: Implications and Strategies," *Translational Neurodegeneration* 6 (2017): 6.

73. "Holdings Report," Protein Data Bank, accessed July 1, 2020, http://www.rcsb .org/pdb/statistics/holdings.do.

74. Matt Reynolds, "DeepMind's AI Is Getting Closer to Its First Big Real-World Application," *Wired*, January 15, 2020, https://www.wired.co.uk/article/deepmind -protein-folding-alphafold.

75. Mohammed AlQuraishi, "AlphaFold at CASP13," *Bioinformatics* 35, no. 22 (2019): 4862–4865.

76. Mohammed AlQuraishi, "AlphaFold @ CASP13: 'What Just Happened?,'" *Some Thoughts on a Mysterious Universe*, December 9, 2018, https://moalquraishi.wordpress .com/2018/12/09/alphafold-casp13-what-just-happened/.

77. Will Douglas Heaven, "DeepMind's Protein-Folding AI Has Solved a 50-Year-Old Grand Challenge of Biology," *MIT Technology Review*, November 30, 2020, https://www.technologyreview.com/2020/11/30/1012712/deepmind-protein-folding -ai-solved-biology-science-drugs-disease/.

78. Heaven, "DeepMind's Protein-Folding AI Has Solved a 50-Year-Old Grand Challenge of Biology"; "AlphaFold: A Solution to a 50-Year-Old Grand Challenge in Biology," DeepMind, November 30, 2020, https://deepmind.com/blog/article /alphafold-a-solution-to-a-50-year-old-grand-challenge-in-biology.

79. Mohammed AlQuraishi, "AlphaFold2 @ CASP14: 'It Feels like One's Child Has Left Home,'" *Some Thoughts on a Mysterious Universe*, December 8, 2020, https:// moalquraishi.wordpress.com/2020/12/08/alphafold2-casp14-it-feels-like-ones-child -has-left-home/.

80. Cade Metz, "London A.I. Lab Claims Breakthrough That Could Accelerate Drug Discovery," *New York Times*, November 30, 2020, https://www.nytimes.com /2020/11/30/technology/deepmind-ai-protein-folding.html.

81. Ewen Callaway, "'It Will Change Everything': DeepMind's AI Makes Gigantic Leap in Solving Protein Structures," *Nature*, November 30, 2020, https://www .nature.com/articles/d41586-020-03348-4.

82. Nicholas Bloom et al., "Are Ideas Getting Harder to Find?," *American Economic Review* 110, no. 4 (April 2020): 1104–1144.

83. Patrick Collison and Michael Nielsen, "Science Is Getting Less Bang for Its Buck," *Atlantic*, November 16, 2018, https://www.theatlantic.com/science/archive /2018/11/diminishing-returns-science/575665/.

84. G. Aad et al., "Combined Measurement of the Higgs Boson Mass in pp Collisions at $\sqrt{s}$=7 and 8 TeV with the ATLAS and CMS Experiments," *Physical Review Letters*, 114, no. 19 (2015): 191803.

85. Collison and Nielsen, "Science Is Getting Less Bang for Its Buck."

86. Robert J. Gordon, *The Rise and Fall of American Growth: The U.S. Standard of Living since the Civil War* (Princeton, NJ: Princeton University Press, 2017).

87. Williams, "Inside DeepMind's Epic Mission to Solve Science's Trickiest Problem"; Richard Jones and James Wilsdon, *The Biomedical Bubble: Why UK Research and Innovation Needs a Greater Diversity of Priorities, Politics, Places and People* (Nesta, July 12, 2018), https://media.nesta.org.uk/documents/The_Biomedical_Bubble_v6.pdf.

88. For an example of machines solving some of the hardest math problems, see Karen Hao, "AI Has Cracked a Key Mathematical Puzzle for Understanding Our World," *MIT Technology Review*, October 30, 2020, https://www.technologyreview

.com/2020/10/30/1011435/ai-fourier-neural-network-cracks-navier-stokes-and -partial-differential-equations.

89. David Pfau et al., "FermiNet: Quantum Physics and Chemistry from First Princi-ples," DeepMind, October 19, 2020, https://deepmind.com/blog/article/FermiNet; Victor Babst et al., "Towards Understanding Glasses with Graph Neural Networks," DeepMind, April 6, 2020, https://deepmind.com/blog/article/Towards-understanding -glasses-with-graph-neural-networks; Gurtej Kanwar et al., "Equivariant Flow-Based Sampling for Lattice Gauge Theory," *Physical Review Letters* 125, no. 12 (September 18, 2020): 121601; David Pfau et al., "Ab Initio Solution of the Many-Electron Schroedinger Equation with Deep Neural Networks," *Physical Review Research* 2, no. 3 (September 16, 2020): 033429.

90. Daniel Oberhaus, "AI Is Throwing Battery Development into Overdrive," *Wired*, October 12, 2020, https://www.wired.com/story/ai-is-throwing-battery-development -into-overdrive; "Facebook and Carnegie Mellon Launch the Open Catalyst Project to Find New Ways to Store Renewable Energy," Facebook AI, October 14, 2020, https://ai.facebook.com/blog/facebook-and-carnegie-mellon-launch-the-open-catalyst -project-to-find-new-ways-to-store-renewable-energy; Marc G. Bellemare et al., "Autonomous Navigation of Stratospheric Balloons Using Reinforcement Learn-ing," *Nature* 588, no. 7836 (December 2020): 77–82; Emile Mathieu and Maximilian Nickel, "Riemannian Continuous Normalizing Flows," *Advances in Neural Informa-tion Processing Systems* 33 (2020), http://proceedings.neurips.cc/paper/2020/hash/1a a3d9c6ce672447e1e5d0f1b5207e85-Abstract.html.

91. Nick Heath, "Google DeepMind Founder Demis Hassabis: Three Truths about AI," *Tech Republic*, September 24, 2018, https://www.techrepublic.com/article /google-deepmind-founder-demis-hassabis-three-truths-about-ai/.

92. Danny Hernandez, "AI and Efficiency," OpenAI, May 5, 2020, https://openai .com/blog/ai-and-efficiency/.

CHAPTER 3

1. Rich Sutton, "The Bitter Lesson," Incomplete Ideas, March 13, 2019, http:// incompleteideas.net/IncIdeas/BitterLesson.html.

2. "Over 50 Years of Moore's Law," Intel, 2015, https://www.intel.com/content/www /us/en/silicon-innovations/moores-law-technology.html; Lee Bell, "What Is Moore's Law? WIRED Explains the Theory That Defined the Tech Industry," *Wired*, August 28, 2016, https://www.wired.co.uk/article/wired-explains-moores-law.

3. Saif M. Khan and Alexander Mann, *AI Chips: What They Are and Why They Matter* (Georgetown University Center for Security and Emerging Technology, April 2020), 13, https://cset.georgetown.edu/wp-content/uploads/AI-Chips%E2%80%94What-They -Are-and-Why-They-Matter.pdf.

4. This is sometimes also known as Rock's Law, after the American investor Arthur Rock.

5. To be more technically precise, the photolithography tool imprints the design into a chemical called a photoresist, which is temporarily coated on the wafer. An

etching tool then permanently etches the design imprinted into the photoresist into the wafer. The photoresist is then removed.

6. Yoshio Nishi and Robert Doering, eds., *Handbook of Semiconductor Manufacturing Technology* (Boca Raton, FL: CRC Press, 2000), 1–3.

7. "The Struggle Over Chips Enters A New Phase," *Economist,* January 21, 2021, https://www.economist.com/leaders/2021/01/23/the-struggle-over-chips-enters-a-new -phase.

8. Sam Shead, "Baidu's Value Took a $1.5 Billion Plunge after Its Chief Scientist Announced He's Leaving," *Business Insider,* March 22, 2017, https://www .businessinsider.com/baidu-value-took-a-15-billion-plunge-after-chief-scientist-andrew -ng-announced-hes-leaving-2017-3.

9. Andrew Ng, "Shaping and Policy Search in Reinforcement Learning" (PhD diss., University of California, Berkeley, 2003).

10. Andrew Ng, "My former student Ian Goodfellow moving to Apple is the big ML news of today. Best of luck @goodfellow_ian, and congrats @tim_cook for getting him!," Twitter, April 6, 2019, 4:09 p.m., https://twitter.com/andrewyng/status /1114304151809224704; Daniela Hernandez, "The Man Behind the Google Brain: Andrew Ng and the Quest for the New AI," *Wired,* July 13, 2005, https://www.wired .com/2013/05/neuro-artificial-intelligence/; Nico Pitney, "Inside the Mind That Built Google Brain: On Life, Creativity, and Failure," *Huffington Post,* May 13, 2015, https://www.huffpost.com/entry/andrew-ng_n_7267682.

11. The focus of this section is on the massive increases in computing power that enabled the training of larger neural networks. There were other notable improvements in the time since Ng's experiment on computer chips, such as progress in the task of executing trained models, often in user devices such as cell phones.

12. George S. Almasi and Allan Gottlieb, *Highly Parallel Computing* (Redwood City, CA: Benjamin-Cummings, 1989).

13. In practice, devising efficient parallel algorithms is more complex than this simple example.

14. Daniela Hernandez, "Now You Can Build Google's $1M Artificial Brain on the Cheap," *Wired,* June 17, 2013, https://www.wired.com/2013/06/andrew-ng/.

15. "The Epi Info Story," Centers for Disease Control and Prevention, September 12, 2016, https://www.cdc.gov/epiinfo/storyall.html.

16. Dean's exact employee number is not known, but most estimates place it at around twenty-five. Gideon Lewis-Kraus, "The Great A.I. Awakening," *New York Times,* December 14, 2016, https://www.nytimes.com/2016/12/14/magazine/the-great-ai -awakening.html.

17. Quoc V. Le et al., "Building High-Level Features Using Large Scale Unsupervised Learning," in *Proceedings of the 29th International Conference on Machine Learning,* ed. John Langford (Madison, WI: Omnipress, 2013), 8595–8598; Lewis-Kraus, "The Great A.I. Awakening."

18. Rajat Raina, Anand Madhavan, and Andrew Ng, "Large-Scale Deep Unsupervised Learning Using Graphics Processors," in *Proceedings of the 26th Annual International Conference on Machine Learning*, ed. Michael L. Littman and Léon Bottou (New York: Association for Computing Machinery, 2009), 873–880.

19. Adam Coates et al., "Deep Learning with COTS HPC Systems," in *Proceedings of the 30th International Conference on Machine Learning*, ed. Sanjoy Dasgupta and David McAllester, 28, no. 3 (2013), 1337–1345; "Researchers Deploy GPUs to Build World's Largest Artificial Neural Network," *Nvidia*, June 17, 2013, https://nvidianews.nvidia .com/news/researchers-deploy-gpus-to-build-world-s-largest-artificial-neural-network.

20. Hernandez, "Now You Can Build Google's $1M Artificial Brain on the Cheap."

21. Krizhevsky, Sutskever, and Hinton, "ImageNet Classification with Deep Convolutional Neural Networks," 2.

22. Jordan Novet, "Why a Deep-Learning Genius Left Google & Joined Chinese Tech Shop Baidu (Interview)," *VentureBeat*, July 30, 2014, https://venturebeat.com /2014/07/30/andrew-ng-baidu/.

23. A tensor is a mathematical object commonly used in calculations.

24. Norman P. Jouppi et al., "In-Datacenter Performance Analysis of a Tensor Processing Unit," in *Proceedings of the 44th Annual International Symposium on Computer Architecture* (New York: Association for Computing Machinery, 2017), 1–12.

25. Cade Metz, "Google's Making Its Own Chips Now. Time for Intel to Freak Out," *Wired*, May 19, 2016, https://www.wired.com/2016/05/googles-making-chips-now -time-intel-freak/.

26. "NVIDIA Launches Revolutionary Volta GPU Platform, Fueling Next Era of AI and High Performance Computing," *Nvidia*, May 10, 2017, https://nvidianews.nvidia .com/news/nvidia-launches-revolutionary-volta-gpu-platform-fueling-next-era-of -ai-and-high-performance-computing; Timothy Prickett Morgan, "Nvidia's Tesla Volta GPU Is the Beast of the Datacenter," *Next Platform*, May 10, 2017, https://www .nextplatform.com/2017/05/10/nvidias-tesla-volta-gpu-beast-datacenter/.

27. Nicole Hemsoth, "First In-Depth Look at Google's New Second-Generation TPU," *Next Platform*, May 17, 2017, https://www.nextplatform.com/2017/05/17/first -depth-look-googles-new-second-generation-tpu/.

28. Jeff Dean and Urs Hölzle, "Build and Train Machine Learning Models on Our New Google Cloud TPUs," Google Blog, May 17, 2017, https://blog.google/topics /google-cloud/google-cloud-offer-tpus-machine-learning/; Hemsoth, "First In-Depth Look at Google's New Second-Generation TPU."

29. Dario Amodei and Danny Hernandez, "AI and Compute," OpenAI, May 16, 2018, https://openai.com/blog/ai-and-compute/.

30. DeepMind, "AlphaStar: The Inside Story," YouTube video, January 24, 2019, 5:11, https://www.youtube.com/watch?v=UuhECwm31dM.

31. "AlphaStar: Mastering the Real-Time Strategy Game StarCraft II," DeepMind, January 24, 2019, https://deepmind.com/blog/article/alphastar-mastering-real-time

-strategy-game-starcraft-ii; Santiago Ontañón et al., "A Survey of Real-Time Strategy Game AI Research and Competition in StarCraft," *IEEE Transactions on Computational Intelligence in AI and Games* 5, no. 4 (2013): 2.

32. Hassabis, "The Power of Self-Learning Systems."

33. For the two-player version of Texas Hold'em, see Chad Holloway, "Czech 'Deep-Stack' v. CMU's 'Libratus' Fight for Top AI Dog Status," *Cardschat*, March 7, 2017, https://web.archive.org/web/20201124034602/https://www.cardschat.com/news /czech-deepstack-v-cmus-libratus-fight-for-top-ai-dog-status-40483/. AI mastered the six-player version of the game in 2019, in substantial part due to an increase in compute capability. Douglas Heaven, "No Limit: AI Poker Bot Is First to Beat Professionals at Multiplayer Game," *Nature*, July 11, 2019, https://www.nature.com /articles/d41586-019-02156-9.

34. Tom Simonite, "Google's AI Declares Galactic War on StarCraft," *Wired*, August 9, 2017, https://www.wired.com/story/googles-ai-declares-galactic-war-on-starcraft-/.

35. "AlphaStar: Mastering the Real-Time Strategy Game StarCraft II."

36. Kyle Wiggers, "DeepMind's AlphaStar Final Beats 99.8% of Human StarCraft 2 Players," *VentureBeat*, October 30, 2019, https://venturebeat.com/2019/10/30 /deepminds-alphastar-final-beats-99-8-of-human-starcraft-2-players/.

37. "AlphaStar: Mastering the Real-Time Strategy Game StarCraft II."

38. "AlphaStar: Mastering the Real-Time Strategy Game StarCraft II."

39. This comparison is for specific models of each type of processor and refers to benchmarks for training speed. See Yu Emma Wang, Gu-Yeon Wei, and David Brooks, "Benchmarking TPU, GPU, and CPU Platforms for Deep Learning," arXiv, July 24, 2019, https://arxiv.org/pdf/1907.10701.pdf.

40. Dan Garisto, "Google AI Beats Top Human Players at Strategy Game StarCraft II," *Nature*, October 30, 2019, https://www.nature.com/articles/d41586-019-03298-6.

41. Ken Wang, "DeepMind Achieved StarCraft II GrandMaster Level, but at What Cost?," Start It Up, January 4, 2020, https://medium.com/swlh/deepmind-achieved -starcraft-ii-grandmaster-level-but-at-what-cost-32891dd990e4.

42. "AlphaStar: Mastering the Real-Time Strategy Game StarCraft II"; Wiggers, "DeepMind's AlphaStar Final Beats 99.8% of Human StarCraft 2 Players."

43. "AlphaStar: Grandmaster Level in StarCraft II Using Multi-Agent Reinforcement Learning," DeepMind, October 30, 2019, https://deepmind.com/blog/article /AlphaStar-Grandmaster-level-in-StarCraft-II-using-multi-agent-reinforcement-learning.

44. "AlphaStar: Mastering the Real-Time Strategy Game StarCraft II."

45. The errant comma in the first sentence is present in the original text. Alec Radford et al., "Better Language Models and Their Implications," OpenAI, February 14, 2019, https://openai.com/blog/better-language-models/.

46. Henrique Pondé et al., "OpenAI Five," OpenAI, June 25, 2018, https://openai.com /blog/openai-five/.

47. OpenAI performed some filtering on Reddit links to collect data from higher-quality websites. Alex Radford et al., "Language Models Are Unsupervised Multitask Learners," OpenAI, February 14, 2019, 3, https://cdn.openai.com/better-language -models/language_models_are_unsupervised_multitask_learners.pdf.

48. For more on how transformers work, see Giuliano Giacaglia, "How Transformers Work," Towards Data Science, March 10, 2019, https://towardsdatascience.com /transformers-141e32e69591; Jay Alammar, "The Illustrated Transformer," GitHub, June 27, 2018, https://jalammar.github.io/illustrated-transformer/.

49. The example given here is from the authors, but for others see Radford et al., "Language Models Are Unsupervised Multitask Learners."

50. The example given here is from the authors, but for others see Radford et al., "Language Models Are Unsupervised Multitask Learners."

51. These parameters included connections between neurons as well as bias terms, of which we omit discussion to keep the text accessible to nontechnical readers. Radford et al., "Language Models Are Unsupervised Multitask Learners," 4–5.

52. Radford et al., "Language Models Are Unsupervised Multitask Learners," 1.

53. The example given here is from the authors.

54. Radford et al., "Language Models Are Unsupervised Multitask Learners," 7.

55. Tom B. Brown et al., "Language Models Are Few-Shot Learners," arXiv, May 28, 2020, 8, http://arxiv.org/abs/2005.14165.

56. Note that we are again omitting discussion of the bias term for the purposes of readability. The 175 billion number technically refers to both weights and biases.

57. OpenAI noted that the 175 billion parameter model required 3,640 petaflops per second for a whole day; a petaflop is equal to one quadrillion floating-point operations, or calculations. For more, see Brown et al., "Language Models Are Few-Shot Learners," 46.

58. Kyle Wiggers, "OpenAI's Massive GPT-3 Model Is Impressive, but Size Isn't Everything," *VentureBeat*, June 1, 2020, https://venturebeat.com/2020/06/01/ai-machine -learning-openai-gpt-3-size-isnt-everything/.

59. Brown et al., "Language Models Are Few-Shot Learners," 9.

60. Nicole Hemsoth, "The Billion Dollar AI Problem That Just Keeps Scaling," *Next Platform*, February 11, 2021, https://www.nextplatform.com/2021/02/11/the -billion-dollar-ai-problem-that-just-keeps-scaling/.

61. Brown et al., "Language Models Are Few-Shot Learners," 28.

62. Brown et al., "Language Models Are Few-Shot Learners," 28. This is an improvement over GPT-2's performance. Sarah Kreps and Miles McCain, "Not Your Father's Bots," *Foreign Affairs*, April 16, 2020, https://www.foreignaffairs.com/articles/2019 -08-02/not-your-fathers-bots.

63. Stanislas Polu and Ilya Sutskever, "Generative Language Modeling for Automated Theorem Proving," arXiv, September 7, 2020, http://arxiv.org/abs/2009.03393.

64. Will Douglas Heaven, "OpenAI's New Language Generator GPT-3 Is Shockingly Good—and Completely Mindless," *MIT Technology Review*, July 20, 2020, https://www.technologyreview.com/2020/07/20/1005454/openai-machine-learning-language-generator-gpt-3-nlp/.

65. Shivkanth Bagavathy, "Introducing GPT3 x GAN/AI generated faces using natural language powered by GPT-3./Eg—'Generate a female face with blonde hair and green eyes,'" September 4, 2020, 5:23 p.m., https://twitter.com/shivkanthb/status/1302039696692789249?s=20; Aditya Ramesh et al., "DALL·E: Creating Images from Text," OpenAI, January 5, 2021, https://openai.com/blog/dall-e/.

66. Scott Alexander, "GPT-2 As Step Toward General Intelligence," *Slate Star Codex*, February 20, 2019, https://slatestarcodex.com/2019/02/19/gpt-2-as-step-toward-general-intelligence/.

67. Wiggers, "OpenAI's Massive GPT-3 Model Is Impressive, but Size Isn't Everything"; Gary Marcus and Ernest Davis, "GPT-3, Bloviator: OpenAI's Language Generator Has No Idea What It's Talking about," *MIT Technology Review*, August 22, 2020, https://www.technologyreview.com/2020/08/22/1007539/gpt3-openai-language-generator-artificial-intelligence-ai-opinion/.

68. Alexander, "GPT-2 As Step Toward General Intelligence."

69. Tekla S. Perry, "Morris Chang: Foundry Father," *IEEE Spectrum*, April 19, 2011, https://spectrum.ieee.org/at-work/tech-careers/morris-chang-foundry-father.

70. "Oral History Interview: Morris Chang," SEMI, August 24, 2007, https://www.semi.org/en/Oral-History-Interview-Morris-Chang.

71. Perry, "Morris Chang: Foundry Father"; Boss Dai, "The Sour Past of 'China Chips,'" trans. Jeffrey Ding and Lorand Laskai, unpublished Google Doc, accessed July 1, 2020, 13, https://docs.google.com/document/d/1g2B0YEFNkPilLB6PTGOQ0NxaO6CflWxvI4zH27h3ISw/edit.

72. Perry, "Morris Chang: Foundry Father."

73. "Oral History Interview: Morris Chang"; Perry, "Morris Chang: Foundry Father."

74. Perry, "Morris Chang: Foundry Father."

75. "Oral History Interview: Morris Chang."

76. Perry, "Morris Chang: Foundry Father."

77. "Oral History Interview: Morris Chang"; Perry, "Morris Chang: Foundry Father."

78. "Oral History Interview: Morris Chang."

79. Perry, "Morris Chang: Foundry Father."

80. Perry, "Morris Chang: Foundry Father"; "Oral History Interview: Morris Chang."

81. Perry, "Morris Chang: Foundry Father."

82. Perry, "Morris Chang: Foundry Father."

83. "Oral History Interview: Morris Chang."

84. Perry, "Morris Chang: Foundry Father."

85. Perry, "Morris Chang: Foundry Father."

86. Perry, "Morris Chang: Foundry Father."

87. "Oral History Interview: Morris Chang"; Perry, "Morris Chang: Foundry Father."

88. Perry, "Morris Chang: Foundry Father."

89. Perry, "Morris Chang: Foundry Father."

90. Perry, "Morris Chang: Foundry Father." This count includes nine TSMC fabs in Taiwan, two fabs in wholly owned subsidiaries (one in China and one in the United States), as well as a joint-venture fab in Singapore.

91. "History of Taiwan Semiconductor Manufacturing Company Ltd," Funding Universe, accessed July 1, 2020, http://www.fundinguniverse.com/company-histories/taiwan-semiconductor-manufacturing-company-ltd-history/.

92. Kim Lyons, "Apple Supplier TSMC Confirms It's Building an Arizona Chip Plant," *Verge*, May 14, 2020, https://www.theverge.com/2020/5/14/21259094/apple-tsmc-factory-chips-arizona-a-series.

93. "TSMC Says Latest Chip Plant Will Cost around $20 Bln," *Reuters*, December 7, 2017, https://www.reuters.com/article/tsmc-investment-idUSL3N1O737Z.

94. Tim Culpan, "The Godfather's 7,300% Return Is Quite the Legacy," *Bloomberg*, June 4, 2018, https://www.bloomberg.com/opinion/articles/2018-06-04/a-7-300-return-for-the-godfather-is-quite-a-legacy.

95. "Oral History Interview: Morris Chang."

96. Dai, "The Sour Past of 'China Chips,'" 5, 9.

97. Paul Mozur and Cade Metz, "A U.S. Secret Weapon in A.I.: Chinese Talent," *New York Times*, June 9, 2020, https://www.nytimes.com/2020/06/09/technology/china-ai-research-education.html.

98. Dai, "The Sour Past of 'China Chips,'" 6–7; Lorand Laskai, "Why Blacklisting Huawei Could Backfire," *Foreign Affairs*, June 21, 2019, https://www.foreignaffairs.com/articles/china/2019-06-19/why-blacklisting-huawei-could-backfire.

99. "China Integrated Circuit Industry Investment Fund (CICF)," Crunchbase, accessed July 1, 2020, https://www.crunchbase.com/organization/china-integrated-circuit-industry-investment-fund-cicf; Li Tao, "How China's 'Big Fund' Is Helping the Country Catch up in the Global Semiconductor Race," *South China Morning Post*, May 10, 2018, https://www.scmp.com/tech/enterprises/article/2145422/how-chinas-big-fund-helping-country-catch-global-semiconductor-race.

100. Will Knight, "China Has Never Had a Real Chip Industry. Making AI Chips Could Change That," *MIT Technology Review*, December 14, 2018, https://www.technologyreview.com/2018/12/14/138260/china-has-never-had-a-real-chip-industry-making-ai-chips-could-change-that/; Allen, *Understanding China's AI Strategy*, 18.

101. Graham Webster et al., "Full Translation: China's 'New Generation Artificial Intelligence Development Plan' (2017)," New America, August 1, 2017, https://www.newamerica.org/cybersecurity-initiative/digichina/blog/full-translation-chinas-new-generation-artificial-intelligence-development-plan-2017/.

102. Paul Mozur, "Inside a Heist of American Chip Designs, as China Bids for Tech Power," *New York Times*, June 22, 2018, https://www.nytimes.com/2018/06/22 /technology/china-micron-chips-theft.html.

103. Josh Ye, "TSMC founder Morris Chang says China's semiconductor industry still five years behind despite decades of subsidies," *South China Morning Post*, April 22, 2021, https://www.scmp.com/tech/big-tech/article/3130628/tsmc-founder-morris -chang-says-chinas-semiconductor-industry-still.

104. Brad Glosserman, "Micro but Mighty: Semiconductors Remain the Key to Technology Leadership," *Japan Times*, May 8, 2020, https://www.japantimes.co.jp /opinion/2020/05/08/commentary/world-commentary/micro-mighty-semiconductors -remain-key-technology-leadership/; James Andrew Lewis, *China's Pursuit of Semiconductor Independence* (Center for Security and International Studies, February 27, 2019), https://www.csis.org/analysis/chinas-pursuit-semiconductor-independence; Bill McClean, "China Forecast to Fall Far Short of Its 'Made in China 2025' Goals for ICs," *IC Insights*, January 6, 2021, https://www.icinsights.com/data/articles /documents/1330.pdf.

105. Saif M. Khan, *Maintaining the AI Chip Competitive Advantage of the United States and Its Allies* (Georgetown University Center for Security and Emerging Technology, December 2019), 3–4, https://cset.georgetown.edu/research/maintaining-the-ai-chip -competitive-advantage-of-the-united-states-and-its-allies/; Khan and Mann, "AI Chips: What They Are and Why They Matter," 7.

106. "China to Import $300 Billion of Chips for Third Straight Year: Industry Group," *Reuters*, August 26, 2020, https://www.reuters.com/article/us-china-semiconductors -idUSKBN25M1CX; Alex Webb, "China's $150 Billion Chip Push Has Hit a Dutch Snag," *Bloomberg*, January 21, 2021, https://www.bloomberg.com/opinion/articles /2021-01-20/asml-china-s-150-billion-chip-push-has-hit-a-dutch-snag.

CHAPTER 4

1. Technically speaking, these were 250 "lexical items."

2. John Hutchins, "The First Public Demonstration of Machine Translation: The Georgetown-IBM System, 7th January 1954" (2006): 34, https://pdfs.semanticscholar .org/ad01/4a7b7a3142e6f17eedddf4218489b56ab18e.pdf.

3. Hutchins, "The First Public Demonstration of Machine Translation," 7.

4. Hutchins, "The First Public Demonstration of Machine Translation," 4–5.

5. Hutchins, "The First Public Demonstration of Machine Translation," 24–25.

6. For a good overview of this era and the limitations of such optimism, see Melanie Mitchell, "Why AI Is Harder Than We Think," arXiv, April 28, 2021, https://arxiv. org/pdf/2104.12871.pdf.

7. Hutchins, "The First Public Demonstration of Machine Translation," 25–27.

8. John Hutchins, "'The Whisky Was Invisible', or Persistent Myths of MT," *MT News International*, June 1995, https://web.archive.org/web/20200213053425/http://www .hutchinsweb.me.uk/MTNI-11-1995.pdf.

9. J. R. Pierce, *Language and Machines: Computers in Translation and Linguistics* (National Research Council, 1966), 32, https://web.archive.org/web/20200522023920 /http://www.mt-archive.info/ALPAC-1966.pdf.

10. John Hutchins, "ALPAC: The (In)Famous Report," Hutchins Web, June 1996, 7, https://web.archive.org/web/20201003194209/http://www.hutchinsweb.me .uk/MTNI-14-1996.pdf; Thierry Poibeau, "The 1966 ALPAC Report and Its Consequences," in *Machine Translation* (Cambridge, MA: MIT Press, 2017), 75–89.

11. There are some exceptions, such as "An Understanding of AI's Limitations Is Starting to Sink In," *Economist*, June 11, 2020, https://www.economist.com /technology-quarterly/2020/06/11/an-understanding-of-ais-limitations-is-starting -to-sink-in; Gary Marcus and Ernest Davis, *Rebooting AI: Building Artificial Intelligence We Can Trust* (New York: Vintage, 2019).

12. Andrew J. Lohn, "Estimating the Brittleness of AI: Safety Integrity Levels and the Need for Testing Out-Of-Distribution Performance," arXiv, September 2, 2020, http://arxiv.org/abs/2009.00802; Michèle A. Flournoy, Avril Haines, and Gabrielle Chefitz, *Building Trust through Testing* (Georgetown University Center for Security and Emerging Technology, October 2020), https://cset.georgetown.edu/wp-content /uploads/Building-Trust-Through-Testing.pdf.

13. Matthew Hutson, "Eye-Catching Advances in Some AI Fields Are Not Real," *Science*, May 27, 2020, https://www.sciencemag.org/news/2020/05/eye-catching-advances -some-ai-fields-are-not-real.

14. Jeffrey Dastin, "Amazon Scraps Secret AI Recruiting Tool That Showed Bias against Women," *Reuters*, October 10, 2018, https://www.reuters.com/article/us-amazon -com-jobs-automation-insight-idUSKCN1MK08G.

15. "Amazon to Create More Than 100,000 New, Full-Time, Full-Benefit Jobs across the U.S. over the Next 18 Months," *Business Wire*, January 12, 2017, https://www .businesswire.com/news/home/20170112005428/en/Amazon-Create-100000-New -Full-Time-Full-Benefit-Jobs.

16. Dastin, "Amazon Scraps Secret AI Recruiting Tool That Showed Bias against Women."

17. Dave Gershgorn, "Companies Are on the Hook If Their Hiring Algorithms Are Biased," *Quartz*, October 22, 2018, https://qz.com/1427621/companies-are-on-the -hook-if-their-hiring-algorithms-are-biased/.

18. Reuters reported in 2018 that the company disbanded the team "by the start of last year." Dastin, "Amazon Scraps Secret AI Recruiting Tool That Showed Bias against Women."

19. Dastin, "Amazon Scraps Secret AI Recruiting Tool That Showed Bias against Women."

20. Katyanna Quach, "MIT Apologizes, Permanently Pulls Offline Huge Dataset That Taught AI Systems to Use Racist, Misogynistic Slurs," *Register*, July 1, 2020, https:// www.theregister.com/2020/07/01/mit_dataset_removed/.

21. Timnit Gebru et al., "Datasheets for Datasets," arXiv, March 23, 2018, http:// arxiv.org/abs/1803.09010; Eun Seo Jo and Timnit Gebru, "Lessons from Archives:

Strategies for Collecting Sociocultural Data in Machine Learning," arXiv, December 22, 2019, http://arxiv.org/abs/1912.10389; Margaret Mitchell et al., "Model Cards for Model Reporting," arXiv, October 5, 2018, http://arxiv.org/abs/1810.03993.

22. Hu Han and Anil K. Jain, "Age, Gender and Race Estimation from Unconstrained Face Images," *MSU Technical Report* (2014): MSU-CSE-14-5, http://biometrics.cse .msu.edu/Publications/Face/HanJain_UnconstrainedAgeGenderRaceEstimation _MSUTechReport2014.pdf; Yaniv Taigman et al., "Deepface: Closing the Gap to Human-Level Performance in Face Verification," in *Proceedings of the IEEE Conference on Computer Vision and Pattern Recognition* (Columbus, OH: IEEE, 2014), 1701–1708. Bias can also affect unsupervised learning systems. Ryan Steed and Aylin Caliskan, "Image Representations Learned With Unsupervised Pre-Training Contain Human-like Biases," arXiv, October 28, 2020, http://arxiv.org/abs/2010.15052.

23. Karen Hao, "We Read the Paper That Forced Timnit Gebru out of Google. Here's What It Says," *MIT Technology Review*, December 4, 2020, https://www .technologyreview.com/2020/12/04/1013294/google-ai-ethics-research-paper-forced -out-timnit-gebru/.

24. Sarah M. West, Meredith Whittaker, and Kate Crawford, *Discriminating Systems: Gender, Race, and Power in AI* (AI Now Institute, April 2019), 6, https://ainowinstitute .org/discriminatingsystems.pdf; Gina Neff, "Does AI Have Gender?," Oxford Internet Institute, recorded June 25, 2018, 23:00–23:45, https://www.oii.ox.ac.uk/videos /does-ai-have-gender/; Joy Buolamwini, "Face: The Final Frontier of Privacy—Full Spoken Congressional Testimony May 22, 2019," Medium, May 23, 2019, https:// medium.com/@Joy.Buolamwini/face-the-final-frontier-of-privacy-full-spoken -congressional-testimony-may-22-2019-ff8607df045b; Virginia Eubanks, *Automating Inequality: How High-Tech Tools Profile, Police, and Punish the Poor* (New York: Picador, 2018), 84–127.

25. Jackie Snow, "'We're in a Diversity Crisis': Cofounder of Black in AI on What's Poisoning Algorithms in Our Lives," *MIT Technology Review*, February 14, 2018, https://www.technologyreview.com/2018/02/14/145462/were-in-a-diversity-crisis -black-in-ais-founder-on-whats-poisoning-the-algorithms-in-our/; Jessi Hempel et al., "Melinda Gates and Fei-Fei Li Want to Liberate AI from 'Guys With Hoodies,'" *Wired*, May 4, 2017, https://www.wired.com/2017/05/melinda-gates-and-fei-fei-li -want-to-liberate-ai-from-guys-with-hoodies/; Karen Hao, "This Is How AI Bias Really Happens—and Why It's So Hard to Fix," *MIT Technology Review*, February 4, 2019, https://www.technologyreview.com/2019/02/04/137602/this-is-how-ai-bias-really -happensand-why-its-so-hard-to-fix/; Bogdana Rakova et al., "Where Responsible AI Meets Reality: Practitioner Perspectives on Enablers for Shifting Organizational Practices," arXiv, June 22, 2020, http://arxiv.org/abs/2006.12358.

26. Richard van Noorden, "The Ethical Questions That Haunt Facial-Recognition Research," *Nature* 587 (November 18, 2020): 354–358, https://www.nature.com /articles/d41586-020-03187-3.

27. Patrick Grother, Mei Ngan, and Kayee Hanaoka, *Face Recognition Vendor Test: Part 3, Demographic Effects* (National Institute of Standards and Technology, December 2019), https://nvlpubs.nist.gov/nistpubs/ir/2019/NIST.IR.8280.pdf.

28. Joy Buolamwini and Timnit Gebru, "Gender Shades: Intersectional Accuracy Disparities in Commercial Gender Classification," in *Proceedings of the 1st Conference on Fairness, Accountability and Transparency*, ed. Sorelle A. Friedler and Christo Wilson, vol. 81, Proceedings of Machine Learning Research (New York: PMLR, 2018), 77–91; Joy Buolamwini, "Written Testimony," Hearing on Facial Recognition Technology (Part 1): Its Impact on Our Civil Rights and Liberties (United States House Committee on Oversight and Government Reform, May 22, 2019), https://docs.house .gov/meetings/GO/GO00/20190522/109521/HHRG-116-GO00-Wstate-BuolamwiniJ -20190522.pdf; Jacob Snow, "Amazon's Face Recognition Falsely Matched 28 Members of Congress With Mugshots," American Civil Liberties Union, July 26, 2018, https://www.aclu.org/blog/privacy-technology/surveillance-technologies/amazons -face-recognition-falsely-matched-28.

29. Bobby Allyn, "'The Computer Got It Wrong': How Facial Recognition Led to False Arrest of Black Man," *NPR*, June 24, 2020, https://www.npr.org/2020/06/24 /882683463/the-computer-got-it-wrong-how-facial-recognition-led-to-a-false-arrest -in-michig; Kashmir Hill, "Another Arrest, and Jail Time, Due to a Bad Facial Recognition Match," *New York Times*, December 29, 2020, https://www.nytimes.com /2020/12/29/technology/facial-recognition-misidentify-jail.html; see also "New Zealand Passport Robot Tells Applicant of Asian Descent to Open Eyes," *Reuters*, December 7, 2016, https://www.reuters.com/article/us-newzealand-passport-error -idUSKBN13W0RL. Others worry that potentially biased algorithms will entrench poverty. Karen Hao, "The Coming War on the Hidden Algorithms That Trap People in Poverty," *MIT Technology Review*, December 4, 2020, https://www.technologyreview .com/2020/12/04/1013068/algorithms-create-a-poverty-trap-lawyers-fight-back/.

30. Jeremy C. Fox, "Brown University Student Mistakenly Identified as Sri Lanka Bombing Suspect," *Boston Globe*, April 28, 2019, https://www.bostonglobe.com /metro/2019/04/28/brown-student-mistaken-identified-sri-lanka-bombings-suspect /0hP2YwyYi4qrCEdxKZCpZM/story.html; Buolamwini, "Written Testimony."

31. Ziad Obermeyer et al., "Dissecting Racial Bias in an Algorithm Used to Manage the Health of Populations," *Science* 366, no. 6464 (October 25, 2019): 447–453. See also Darshali A. Vyas, Leo G. Eisenstein, and David S. Jones, "Hidden in Plain Sight— Reconsidering the Use of Race Correction in Clinical Algorithms," *New England Journal of Medicine* 383 (August 27, 2020), https://www.nejm.org/doi/full/10.1056 /NEJMms2004740.

32. Amanda Askell, "In AI Ethics, 'Bad' Isn't Good Enough," Askell.io, December 14, 2020, https://askell.io/posts/2020/12/bad-isnt-good-enough.

33. Pedro Domingos, "True fact: there is not a single proven case to date of discrimination by machine learning algorithms," Twitter, September 20, 2017, 11:20 p.m., https://twitter.com/pmddomingos/status/910388223142010880.

34. For an early articulation of a similar idea, see Maciej Cegłowski, "The Moral Economy of Tech," talk given at Society for the Advancement of Socio-Economics, June 26, 2016, https://idlewords.com/talks/sase_panel.htm.

35. Pratyusha Kalluri, "Don't Ask If Artificial Intelligence Is Good or Fair, Ask How It Shifts Power," *Nature* 583, no. 7815 (July 9, 2020): 169; Frank Pasquale, "The Second Wave of Algorithmic Accountability," Law and Political Economy Project, November 25, 2019, https://lpeproject.org/blog/the-second-wave-of-algorithmic -accountability/.

36. For an example of algorithms increasing fairness, see Emma Pierson et al., "An Algorithmic Approach to Reducing Unexplained Pain Disparities in Underserved Populations," *Nature Medicine* 27, no. 1 (2021): 136–140.

37. Note that Hopper did not invent the term *bug*, but she helped popularize it. "Log Book With Computer Bug," National Museum of American History, accessed July 1, 2020, https://americanhistory.si.edu/collections/search/object/nmah_334663; Sharron Ann Danis, "Rear Admiral Grace Murray Hopper," February 16, 1997, http:// ei.cs.vt.edu/~history/Hopper.Danis.html; Alexander Magoun and Paul Israel, "Did You Know? Edison Coined the Term 'Bug,'" *Institute*, August 23, 2013, https://web .archive.org/web/20160304130915/http://theinstitute.ieee.org/technology-focus /technology-history/did-you-know-edison-coined-the-term-bug.

38. Some long-shot technology efforts, such as DARPA's Explainable AI project, aim to remedy this. For more, see Matt Turek, "Explainable Artificial Intelligence (XAI)," DARPA, accessed July 1, 2020, https://www.darpa.mil/program/explainable -artificial-intelligence.

39. Campbell originally said "man" rather than "human," leading to this excellent response: Nicole Rudick, "A Universe of One's Own," *New York Review of Books*, July 18, 2019, https://www.nybooks.com/articles/2019/07/18/universe-of-ones-own-women -science-fiction/.

40. Iyad Rahwan et al., "Machine Behaviour," *Nature* 568, no. 7753 (2019): 477–486.

41. Ron Schmelzer, "Towards A More Transparent AI," *Forbes*, May 23, 2020, https:// www.forbes.com/sites/cognitiveworld/2020/05/23/towards-a-more-transparent-ai/; Stefan Larsson and Fredrik Heintz, "Transparency in Artificial Intelligence," *Internet Policy Review* 9, no. 2 (2020), https://policyreview.info/concepts/transparency -artificial-intelligence.

42. This erratic behavior has led to concerns about researchers cherry-picking results and a reproducibility crisis in AI. Elizabeth Gibney, "This AI Researcher Is Trying to Ward Off a Reproducibility Crisis," *Nature* 577, no. 7788 (December 2019), https:// www.ncbi.nlm.nih.gov/pubmed/31871325.

43. Robert O. Work and Paul Scharre, "Transcript from Military AI Applications," Center for a New American Security, April 29, 2020, https://www.cnas.org /publications/transcript/transcript-from-military-ai-applications.

44. Paul Robinette et al., "Overtrust of Robots in Emergency Evacuation Scenarios," in *2016 11th ACM/IEEE International Conference on Human-Robot Interaction* (2016), 101–108, https://www.cc.gatech.edu/~alanwags/pubs/Robinette-HRI-2016.pdf.

45. Bryce Goodman and Seth Flaxman, "European Union Regulations on Algorithmic Decision-Making and a 'Right to Explanation,'" *AI Magazine* 38, no. 3 (2017):

50–57; Sandra Wachter, Brent Mittelstadt, and Luciano Floridi, "Why a Right to Explanation of Automated Decision-Making Does Not Exist in the General Data Protection Regulation," *International Data Privacy Law* 7, no. 2 (2017): 76–99.

46. Thomas Elsken, Jan Hendrik Metzen, and Frank Hutter, "Neural Architecture Search: A Survey," *Journal of Machine Learning Research*, August 16, 2018, http://arxiv.org/abs/1808.05377; Barret Zoph and Quoc V. Le, "Neural Architecture Search with Reinforcement Learning," in *5th International Conference on Learning Representations* (2016), http://arxiv.org/abs/1611.01578.

47. Dario Amodei and Jack Clark, "Faulty Reward Functions in the Wild," OpenAI, December 21, 2016, https://openai.com/blog/faulty-reward-functions/.

48. DeepMind, "Specification Gaming Examples in AI—Master List," Google Sheet, accessed April 21, 2021, https://docs.google.com/spreadsheets/d/e/2PACX-1vRPiprOaC3HsCf5Tuum8bRfzYUiKLRqJmbOoC-32JorNdfyTiRRsR7Ea5eWtvs Wzuxo8bjOxCG84dAg/pubhtml; Victoria Krakovna et al., "Specification Gaming: The Flip Side of AI Ingenuity," DeepMind, April 21, 2020, https://deepmind.com/blog/article/Specification-gaming-the-flip-side-of-AI-ingenuity.

49. Neel V. Patel, "Why Doctors Aren't Afraid of Better, More Efficient AI Diagnosing Cancer," *Daily Beast*, December 11, 2017, https://www.thedailybeast.com/why-doctors-arent-afraid-of-better-more-efficient-ai-diagnosing-cancer; Lisette Hilton, "The Artificial Brain as Doctor," *Medpage Today*, January 15, 2018, https://www.medpagetoday.com/dermatology/generaldermatology/70513; John R. Zech et al., "Variable Generalization Performance of a Deep Learning Model to Detect Pneumonia in Chest Radiographs: A Cross-Sectional Study," *PLoS Medicine* 15, no. 11 (2018), https://doi.org/10.1371/journal.pmed.1002683.

50. Tom Everitt and Marcus Hutter, "Reward Tampering Problems and Solutions in Reinforcement Learning: A Causal Influence Diagram Perspective," arXiv, August 13, 2019, http://arxiv.org/abs/1908.04734; Tom Everitt, Ramana Kumar, and Marcus Hutter, "Designing Agent Incentives to Avoid Reward Tampering," *DeepMind Safety Research*, August 14, 2019, https://medium.com/@deepmindsafetyresearch/designing-agent-incentives-to-avoid-reward-tampering-4380c1bb6cd.

51. Dario Amodei et al., "Concrete Problems in AI Safety," arXiv, June 21, 2016, 3–4, http://arxiv.org/abs/1606.06565; Heather Roff, "AI Deception: When Your Artificial Intelligence Learns to Lie," *IEEE Spectrum*, February 24, 2020, https://spectrum.ieee.org/automaton/artificial-intelligence/embedded-ai/ai-deception-when-your-ai-learns-to-lie.

52. This is known as Goodhart's Law. "Goodhart's Law," Oxford Reference, accessed July 1, 2020 https://www.oxfordreference.com/view/10.1093/oi/authority.20110803095859655; Jonathan Law, *A Dictionary of Finance and Banking* (Oxford: Oxford University Press, 2014).

53. Stephen J. Dubner, "The Cobra Effect (Ep. 96)," October 11, 2012, *Freakonomics*, podcast, https://freakonomics.com/podcast/the-cobra-effect-a-new-freakonomics-radio-podcast/.

54. Amodei et al., "Concrete Problems in AI Safety," 7.

55. Pedro A. Ortega and Vishal Maini, "Building Safe Artificial Intelligence: Specification, Robustness, and Assurance," *DeepMind Safety Research*, September 27, 2018, https://medium.com/@deepmindsafetyresearch/building-safe-artificial-intelligence-52f5f75058f1.

56. Amodei et al., "Concrete Problems in AI Safety," 8.

57. John R. Searle, "Minds, Brains, and Programs," *Behavioral and Brain Sciences* 3, no. 3 (1980): 417–424; David Cole, "The Chinese Room Argument," *Stanford Encyclopedia of Philosophy* (2020), https://plato.stanford.edu/archives/spr2020/entries/chinese-room/.

58. Edsger W. Dijkstra, "The Threats to Computing Science," talk delivered at the ACM 1984 South Central Regional Conference, November 16–18, 1984, University of Texas, Austin, https://www.cs.utexas.edu/users/EWD/transcriptions/EWD08xx/EWD898.html.

59. Stuart J. Russell and Peter Norvig, *Artificial Intelligence: A Modern Approach*, 2nd ed. (Upper Saddle River, NJ: Prentice Hall, 2003), 947.

60. John Seabrook, "Can a Machine Learn to Write for The New Yorker?," *New Yorker*, October 7, 2019, https://www.newyorker.com/magazine/2019/10/14/can-a-machine-learn-to-write-for-the-new-yorker.

61. Seabrook, "Can a Machine Learn to Write for The New Yorker?" Microsoft has a major research effort to build one trillion parameter neural networks. Deepak Narayanan, "Efficient Large-Scale Language Model Training on GPU Clusters," arXiv, April 9, 2021, https://arxiv.org/abs/2104.04473. Note that, in 2021, Facebook revealed multitrillion parameter neural networks, but these were of a different type than the generative models used in GPT-3 and should not be compared directly.

CHAPTER 5

1. For the definitive history, see Richard Rhodes, *The Making of the Atomic Bomb* (New York: Simon & Schuster, 1986).

2. Rhodes, *The Making of the Atomic Bomb,* 664; Lansing Lamont, *Day of Trinity* (New York: Scribner, 1965), 197.

3. "Ralph Smith's Eyewitness Account of the Trinity Trip to Watch Blast," White Sands Missile Range, New Mexico (US Army), December 4, 2010, https://web.archive.org/web/20140904084107/http://www.wsmr.army.mil/PAO/Trinity/Pages/RalphSmithseyewitnessaccountoftheTrinitytriptowatchblast.aspx.

4. Alex Wellerstein, "The First Light of Trinity," *New Yorker*, July 16, 2015, https://www.newyorker.com/tech/annals-of-technology/the-first-light-of-the-trinity-atomic-test.

5. Kenneth T. Bainbridge, "'All in Our Time'—A Foul and Awesome Display," *Bulletin of the Atomic Scientists* 31, no. 5 (1975): 46.

6. "Videos: Footage of Historical Events and Nuclear Weapons Effects," Atomic Archive, accessed July 1, 2020, https://www.atomicarchive.com/media/videos/index.html.

7. James Temperton, "'Now I Am Become Death, the Destroyer of Worlds'? The Story of Oppenheimer's Infamous Quote," *Wired*, August 9, 2017, https://www.wired .co.uk/article/manhattan-project-robert-oppenheimer.

8. "Manhattan Project: The Manhattan Project and the Second World War (1939–1945)," Department of Energy, accessed July 1, 2020, https://www.osti.gov/opennet /manhattan-project-history/Events/1945/retrospect.htm.

9. Robert Reinhold, "Dr. Vannevar Bush Is Dead at 84," *New York Times*, June 30, 1974, https://www.nytimes.com/1974/06/30/archives/dr-vannevar-bush-is-dead-at -84-dr-vannevar-bush-who-marshaled.html.

10. Jerome B. Wiesner, *Vannevar Bush, 1890–1974* (National Academy of Science, 1979), 91, http://www.nasonline.org/publications/biographical-memoirs/memoir-pdfs /bush-vannevar.pdf.

11. G. Pascal Zachary, *Endless Frontier: Vannevar Bush, Engineer of the American Century* (New York: Free Press, 1997), 28–32.

12. Claude E. Shannon, "Claude E. Shannon, an Oral History Conducted in 1982 by Robert Price" (IEEE History Center, Piscataway, NJ), Engineering Technology and History Wiki, accessed July 1, 2020, https://ethw.org/Oral-History:Claude_E._Shannon.

13. Walter Isaacson, "How America Risks Losing Its Innovation Edge," *Time*, January 3, 2019, https://time.com/longform/america-innovation/.

14. Zachary, *Endless Frontier*, 104–112.

15. Jeremy Hsu, "This Tycoon's Secret Radar Lab Helped Win WWII," *Discover Magazine*, January 16, 2018, https://www.discovermagazine.com/technology/this -tycoons-secret-radar-lab-helped-win-wwii.

16. Jon Gertner, "How Bell Lab's Radar Dismantled the Axis Powers," *Gizmodo*, March 29, 2012, https://gizmodo.com/how-bell-labs-radar-dismantled-the-axis-powers -5897137; Irvin Stewart, *Organizing Scientific Research for War: The Administrative History of the Office of Scientific Research and Development* (Boston: Little, Brown, 1948), 190.

17. Vannevar Bush, "As We May Think," *Atlantic*, July 1, 1945, https://www .theatlantic.com/magazine/archive/1945/07/as-we-may-think/303881/.

18. Vannevar Bush, *Science: The Endless Frontier* (National Science Foundation, July 1945), https://www.nsf.gov/about/history/vbush1945.htm.

19. These quotes originated with Bush's book, *Modern Arms and Free Men*, and are reproduced in his obituary. Reinhold, "Dr. Vannevar Bush Is Dead at 84."

20. Linda Weiss, *America Inc.: Innovation and Enterprise in the National Security State* (Ithaca: Cornell University Press, 2014).

21. Isaacson, "How America Risks Losing Its Innovation Edge."

22. James G. Hershberg, *James B. Conant: Harvard to Hiroshima and the Making of the Nuclear Age* (New York: Knopf, 1993), 305–309.

23. Daniel Holbrook, "Government Support of the Semiconductor Industry: Diverse Approaches and Information Flows," *Business and Economic History* 24, no. 2 (1995):

141–146; Eric M. Jones, "Apollo Lunar Surface Journal," NASA, June 5, 2018, https://www.hq.nasa.gov/alsj/ic-pg3.html.

24. Katrina Onstad, "Mr. Robot," *Toronto Life*, January 29, 2018, https://torontolife.com/tech/ai-superstars-google-facebook-apple-studied-guy/.

25. Cade Metz, *Genius Makers: The Mavericks Who Brought AI to Google, Facebook, and the World* (New York: Dutton, 2021), 31.

26. Onstad, "Mr. Robot."

27. Metz, *Genius Makers*, 33.

28. Craig S. Smith, "The Man Who Helped Turn Toronto Into a High-Tech Hotbed," *New York Times*, June 23, 2017, https://www.nytimes.com/2017/06/23/world/canada/the-man-who-helped-turn-toronto-into-a-high-tech-hotbed.html.

29. In 1958, the precursor to neural networks, known as a perceptron, gained attention when Frank Rosenblatt, a scientist at Cornell, posited that intelligence could be mimicked with an array of computational neurons. Other major names in AI, including Marvin Minsky, dismissed the idea—perhaps out of a desire to win some of the contracts for which Rosenblatt was competing.

30. Smith, "The Man Who Helped Turn Toronto Into a High-Tech Hotbed."

31. Geoffrey E. Hinton, Curriculum Vitae, University of Toronto, January 6, 2020, 36, https://www.cs.toronto.edu/~hinton/fullcv.pdf.

32. Smith, "The Man Who Helped Turn Toronto Into a High-Tech Hotbed."

33. Onstad, "Mr. Robot."

34. Onstad, "Mr. Robot."

35. Smith, "The Man Who Helped Turn Toronto Into a High-Tech Hotbed."

36. Sara Sabour, Nicholas Frosst, and Geoffrey E. Hinton, "Dynamic Routing Between Capsules," arXiv, October 26, 2017, https://arxiv.org/abs/1710.09829.

37. Tiernan Ray, "Google DeepMind's Demis Hassabis Is One Relentlessly Curious Public Face of AI," *ZDNet*, May 6, 2019, https://www.zdnet.com/article/googles-demis-hassabis-is-one-relentlessly-curious-public-face-of-ai/; Sam Shead, "The CEO of Google's £400 Million AI Startup Is Going to Meet with Tech Leaders to Discuss Ethics," *Business Insider*, November 5, 2015, https://www.businessinsider.com/the-ceo-of-googles-400-million-ai-startup-is-going-to-meet-with-tech-leaders-2015-11; Cade Metz, "Pentagon Wants Silicon Valley's Help on A.I.," *New York Times*, March 15, 2018, https://www.nytimes.com/2018/03/15/technology/military-artificial-intelligence.html; "An Open Letter: Research Priorities for Robust and Beneficial Artificial Intelligence," Future of Life Institute, July 18, 2018, https://futureoflife.org/ai-open-letter/.

38. Interview with the authors, April 17, 2020.

39. Interview with the authors.

40. Interview with the authors.

41. Interview with the authors.

42. Cade Metz and Kate Conger, "'I Could Solve Most of Your Problems': Eric Schmidt's Pentagon Offensive," *New York Times*, May 2, 2020, https://www.nytimes.com/2020/05/02/technology/eric-schmidt-pentagon-google.html.

43. Interview with the authors.

44. Interview with the authors.

45. Interview with the authors.

46. Interview with the authors.

47. Scott Shane, Cade Metz, and Daisuke Wakabayashi, "How a Pentagon Contract Became an Identity Crisis for Google," *New York Times*, May 30, 2018, https://www.nytimes.com/2018/05/30/technology/google-project-maven-pentagon.html; Kate Conger, "Google Plans Not to Renew Its Contract for Project Maven, a Controversial Pentagon Drone AI Imaging Program," *Gizmodo*, June 1, 2018, https://gizmodo.com/google-plans-not-to-renew-its-contract-for-project-mave-1826488620. Fei-Fei Li did not respond to requests for an interview.

48. Interview with the authors.

49. Interview with the authors, April 15, 2020.

50. Interview with the authors.

51. Interview with the authors.

52. Interview with the authors.

53. "Letter to Google C.E.O," *New York Times*, April 4, 2018, https://static01.nyt.com/files/2018/technology/googleletter.pdf; Scott Shane and Daisuke Wakabayashi, "'The Business of War': Google Employees Protest Work for the Pentagon," *New York Times*, April 4, 2018, https://www.nytimes.com/2018/04/04/technology/google-letter-ceo-pentagon-project.html.

54. Ben Tarnoff, "Tech Workers Versus the Pentagon," *Jacobin*, June 6, 2018, https://jacobinmag.com/2018/06/google-project-maven-military-tech-workers.

55. Tarnoff, "Tech Workers Versus the Pentagon."

56. Kate Conger, "Google Employees Resign in Protest against Pentagon Contract," *Gizmodo*, May 14, 2018, https://gizmodo.com/google-employees-resign-in-protest-against-pentagon-con-1825729300.

57. Tarnoff, "Tech Workers Versus the Pentagon."

58. Tarnoff, "Tech Workers Versus the Pentagon."

59. Interview with the authors.

60. Shane and Wakabayashi, "'The Business of War.'"

61. Patrick Tucker, "Google Is Pursuing the Pentagon's Giant Cloud Contract Quietly, Fearing an Employee Revolt," *DefenseOne*, April 12, 2018, https://www.defenseone.com/technology/2018/04/google-pursuing-pentagons-giant-cloud-contract-quietly-fearing-employee-revolt/147407/.

62. Tarnoff, "Tech Workers Versus the Pentagon."

63. Kate Conger, "Tech Workers Ask Google, Microsoft, IBM, and Amazon to Not Pursue Pentagon Contracts," *Gizmodo*, April 19, 2018, https://gizmodo.com/tech -workers-ask-google-microsoft-ibm-and-amazon-to-1825394821.

64. "Open Letter in Support of Google Employees and Tech Workers," International Committee for Robot Arms Control, May 1, 2018, https://www.icrac.net/open -letter-in-support-of-google-employees-and-tech-workers/.

65. Erin Griffin, "Google Won't Renew Controversial Pentagon AI Project," *Wired*, June 1, 2018, https://www.wired.com/story/google-wont-renew-controversial-pentagon -ai-project/.

66. Sundar Pichai, "AI at Google: Our Principles," Google, June 7, 2018, https:// www.blog.google/technology/ai/ai-principles/.

67. Interview with the authors.

68. Kate Conger, "Google Removes 'Don't Be Evil' Clause from Its Code of Conduct," *Gizmodo*, May 18, 2018, https://gizmodo.com/google-removes-nearly-all-mentions -of-dont-be-evil-from-1826153393; "Google and YouTube Will Pay Record $170 Million for Alleged Violations of Children's Privacy Law," Federal Trade Commission, September 4, 2019, https://www.ftc.gov/news-events/press-releases/2019/09/google -youtube-will-pay-record-170-million-alleged-violations; Ryan Gallagher, "Inside Google's Effort to Develop a Censored Search Engine in China," *Intercept*, August 8, 2018, https://theintercept.com/2018/08/08/google-censorship-china-blacklist/.

69. Interview with the authors.

70. "DOD Adopts Ethical Principles for Artificial Intelligence," Deptartment of Defense, April 24, 2020, https://www.defense.gov/Newsroom/Releases/Release /Article/2091996/dod-adopts-ethical-principles-for-artificial-intelligence/.

71. Interview with the authors.

72. Interview with the authors.

73. For more on this evolving relationship, including insights into how it varies by type of project, see Catherine Aiken, Rebecca Kagan, and Michael Page, *"Cool Projects" or "Expanding the Efficiency of the Murderous American War Machine?"* (Georgetown University Center for Security and Emerging Technology, November 2020), https://cset.georgetown.edu/wp-content/uploads/CSET-Cool-Projects-or-Expanding -the-Efficiency-of-the-Murderous-American-War-Machine.pdf.

74. Interview with the authors; Meredith Whittaker, "Onward! Another #Google-Walkout Goodbye," Medium, July 16, 2019, https://medium.com/@GoogleWalkout /onward-another-googlewalkout-goodbye-b733fa134a7d.

75. Interview with the authors.

76. Sarah Zhang, "China's Artificial-Intelligence Boom," *Atlantic*, February 16, 2017, https://www.theatlantic.com/technology/archive/2017/02/china-artificial-intelligence /516615/.

77. Murdick, Dunham, and Melot, "AI Definitions Affect Policymaking," 4. As the authors note, "the intersection years shifted earlier for computer vision . . . to

2014 (EU) and 2017 (United States)—and later for robotics—to 2017 (EU) and 2019 (United States). In natural language processing . . . on the other hand, 2019 was an intersection year for China and the EU and there was no intersection for China and the United States." By intersection year, the authors mean the year in which China achieved parity.

78. Shu-Ching Jean Chen, "SenseTime: The Faces Behind China's Artificial Intelligence Unicorn," *Forbes*, March 7, 2018, https://www.forbes.com/sites/shuchingjeanchen /2018/03/07/the-faces-behind-chinas-omniscient-video-surveillance-technology/.

79. Chen, "SenseTime."

80. Yi Sun, Xiaogang Wang, and Xiaoou Tang, "Deep Learning Face Representation from Predicting 10,000 Classes," in *Proceedings of the IEEE Conference on Computer Vision and Pattern Recognition* (2014): 1891–1898, http://www.ee.cuhk.edu.hk /~xgwang/papers/sunWTcvpr14.pdf.

81. John Pomfret, "Tiananmen Square Anniversary: China Tried to Erase the Crackdown from Memory. But Its Legacy Lives On," *Washington Post*, May 30, 2019, https:// www.washingtonpost.com/graphics/2019/opinions/global-opinions/tiananmen -square-a-massacre-erased/; Chen, "SenseTime."

82. Chen, "SenseTime."

83. Will Knight, "China's AI Awakening," *MIT Technology Review*, October 10, 2017, https://www.technologyreview.com/2017/10/10/148284/chinas-ai-awakening/.

84. Lulu Yilun Chen, "World's Most Valuable AI Startup Scrambles to Survive Trump's Blacklist," *Bloomberg*, October 28, 2019, https://www.bloomberg.com/news /articles/2019-10-28/on-the-blacklist-china-s-top-ai-startup-tries-to-survive-trump.

85. "CVPR 2017 Open Access Repository," Open Access, accessed July 1, 2020, https://openaccess.thecvf.com/CVPR2017.

86. Sijia Jiang, "China's SenseTime Valued at $4.5 Billion after Alibaba-Led Funding: Sources," *Reuters*, April 9, 2018, https://www.reuters.com/article/us-sensetime-funding -idUSKBN1HG0CI.

87. Allen, *Understanding China's AI Strategy*.

88. Julie Steinberg and Jing Yang, "Hong Kong's SenseTime Considers $1 Billion Capital Raise," *Wall Street Journal*, May 15, 2020, https://www.wsj.com/articles/hong -kongs-sensetime-considers-1-billion-capital-raise-11589561280.

89. "SenseTime and Qualcomm to Collaborate to Drive On-Device Artificial Intelligence," Qualcomm, October 18, 2017, https://www.qualcomm.com/news/releases /2017/10/20/sensetime-and-qualcomm-collaborate-drive-device-artificial-intelligence.

90. Chen, "World's Most Valuable AI Startup Scrambles to Survive Trump's Blacklist."

91. "SenseTime Debuts in Singapore by Signing Memoranda of Understanding with Local Giants NTU, NSCC and Singtel," *Markets Insider*, June 29, 2019, https:// markets.businessinsider.com/news/stocks/sensetime-debuts-in-singapore-by-signing -memoranda-of-understanding-with-local-giants-ntu-nscc-and-singtel-102733

0334#; "Milestones," SenseTime, December 5, 2018, https://web.archive.org/web/20181205103418/https://www.sensetime.com/road.

92. Catherine Shu, "MIT Is Reviewing Its Relationship with AI Startup SenseTime, One of the Chinese Tech Firms Blacklisted by the US," *Tech Crunch*, October 9, 2019, http://techcrunch.com/2019/10/09/mit-is-reviewing-its-relationship-with-ai-startup -sensetime-one-of-the-chinese-tech-firms-blacklisted-by-the-u-s/.

93. Josh Horwitz, "SenseTime: The Billion-Dollar, Alibaba-Backed AI Company That's Quietly Watching Everyone in China," *Quartz*, April 16, 2018, https://qz.com /1248493/sensetime-the-billion-dollar-alibaba-backed-ai-company-thats-quietly -watching-everyone-in-china/.

94. Knight, "China's AI Awakening." For more on Chinese industry, see Douglas B. Fuller, *Paper Tigers, Hidden Dragons: Firms and the Political Economy of China's Techno-logical Development* (Oxford: Oxford University Press, 2016).

95. Ma Si and Zhou Lanxu, "Tech Giants Rally to Authorities' Call," *China Daily*, November 2, 2018, http://www.chinadaily.com.cn/a/201811/02/WS5bdbaddba310 eff3032861e5.html.

96. Jeffrey Ding, "#84: Biometric Recognition White Paper 2019," *ChinAI Newsletter*, March 1, 2020, https://chinai.substack.com/p/chinai-84-biometric-recognition -white; Mark Bergen and David Ramli, "China's Plan for World Domination in AI Isn't So Crazy After All," *Bloomberg*, August 14, 2017, https://www.bloomberg.com/news /articles/2017-08-14/china-s-plan-for-world-domination-in-ai-isn-t-so-crazy-after-all.

97. Different sources provide different interpretations of how direct the ties are between the company and the government. Chen, "World's Most Valuable AI Startup Scrambles to Survive Trump's Blacklist"; Rebecca Fannin, "The List of What Could Go Wrong in China's Tech Boom Just Got a Lot Longer: AI Ban," *Forbes*, October 8, 2019, https://www.forbes.com/sites/rebeccafannin/2019/10/08/the-list-of -what-could-go-wrong-in-chinas-tech-boom-just-got-a-lot-longer/; Chen, "SenseTime."

98. Jeffrey Ding, "#29 Complicit—China's AI Unicorns and the Securitization of Xinjiang," *ChinAI Newsletter*, September 23, 2018, https://chinai.substack.com/p /chinai-newsletter-29-complicit-chinas-ai-unicorns-and-the-securitization-of -xinjiang; Austin Ramzy, "China's Oppression of Muslims in Xinjiang, Explained," *New York Times*, April 14, 2019, https://www.nytimes.com/2021/01/20/world/asia/china -genocide-uighurs-explained.html; John Hudson, "As Tensions with China Grow, Biden Administration Formalizes Genocide Declaration Against Beijing," *Washing-ton Post*, March 30, 2021, https://www.washingtonpost.com/national-security/china -genocide-human-rights-report/2021/03/30/b2fa8312-9193-11eb-9af7-fd0822ae4398 _story.html.

99. Paul Mozur, "One Month, 500,000 Face Scans: How China Is Using A.I. to Profile a Minority," *New York Times*, April 14, 2019, https://www.nytimes.com/2019/04/14 /technology/china-surveillance-artificial-intelligence-racial-profiling.html.

100. Steinberg and Yang, "Hong Kong's SenseTime Considers $1 Billion Capital Raise"; "China's SenseTime Sells Out of Xinjiang Security Joint Venture," *Financial*

Times, April 15, 2019, https://www.ft.com/content/38aa038a-5f4f-11e9-b285-3acd5
d43599e.

101. Department of Commerce, "Addition of Certain Entities to the Entity List,"
Federal Register, October 9, 2019, 4, https://www.federalregister.gov/documents/2019
/10/09/2019-22210/addition-of-certain-entities-to-the-entity-list.

102. Shu, "MIT Is Reviewing Its Relationship with AI Startup SenseTime."

103. Chen, "World's Most Valuable AI Startup Scrambles to Survive Trump's
Blacklist."

104. Tai Ming Cheung, *Fortifying China: The Struggle to Build a Modern Defense Economy* (Ithaca: Cornell University Press, 2013). See also Elsa B. Kania, "In Military-Civil
Fusion, China Is Learning Lessons from the United States and Starting to Innovate,"
Strategy Bridge, August 27, 2019, https://thestrategybridge.org/the-bridge/2019/8/27
/in-military-civil-fusion-china-is-learning-lessons-from-the-united-states-and-starting
-to-innovate; Lorand Laskai, "Civil-Military Fusion: The Missing Link between
China's Technological and Military Rise," Council on Foreign Relations, January
29, 2018, https://www.cfr.org/blog/civil-military-fusion-missing-link-between-chinas
-technological-and-military-rise; Elsa B. Kania and Lorand Laskai, "Myths and
Realities of China's Military-Civil Fusion Strategy," Center for a New American
Security, January 28, 2021, https://www.cnas.org/publications/reports/myths-and
-realities-of-chinas-military-civil-fusion-strategy.

105. Anja Manuel and Kathleen Hicks, "Can China's Military Win the Tech War?,"
Foreign Affairs, July 30, 2020, https://www.foreignaffairs.com/articles/united-states
/2020-07-29/can-chinas-military-win-tech-war.

106. Murray Scot Tanner, "Beijing's New National Intelligence Law: From Defense to
Offense," *Lawfare*, July 20, 2017, https://www.lawfareblog.com/beijings-new-national
-intelligence-law-defense-offense.

107. Wei, "China's Xi Ramps Up Control of Private Sector"; Lingling Wei, "Jack Ma
Makes Ant Offer to Placate Chinese Regulators," *Wall Street Journal*, December
20, 2020, https://www.wsj.com/articles/jack-ma-makes-ant-offer-to-placate-chinese
-regulators-11608479629.

108. Zach Dorfman, "Tech Giants Are Giving China a Vital Edge in Espionage,"
Foreign Policy, December 23, 2020, https://foreignpolicy.com/2020/12/23/china
-tech-giants-process-stolen-data-spy-agencies/.

109. Dorfman, "Tech Giants Are Giving China a Vital Edge in Espionage."

CHAPTER 6

1. Svetlana V. Savranskaya, "New Sources on the Role of Soviet Submarines in the
Cuban Missile Crisis," *Journal of Strategic Studies* 28, no. 2 (2005): 243.

2. Savranskaya, "New Sources on the Role of Soviet Submarines in the Cuban Missile
Crisis," 241; Michael Evans, "The Submarines of October: Chronology," National
Security Archive, accessed July 1, 2020, https://nsarchive2.gwu.edu/NSAEBB
/NSAEBB75/subchron.htm.

3. "The Man Who Saved the World," *Secrets of the Dead*, October 24, 2012, https://www.pbs.org/wnet/secrets/the-man-who-saved-the-world-watch-the-full-episode/905/.

4. Nicola Davis, "Soviet Submarine Officer Who Averted Nuclear War Honoured with Prize," *Guardian*, October 27, 2017, http://www.theguardian.com/science/2017/oct/27/vasili-arkhipov-soviet-submarine-captain-who-averted-nuclear-war-awarded-future-of-life-prize.

5. Robert Krulwich, "You (and Almost Everyone You Know) Owe Your Life to This Man," *National Geographic*, March 25, 2016, https://www.nationalgeographic.com/news/2016/03/you-and-almost-everyone-you-know-owe-your-life-to-this-man/; Davis, "Soviet Submarine Officer Who Averted Nuclear War Honoured with Prize."

6. Ronald Arkin, "Lethal Autonomous Systems and the Plight of the Non-Combatant," in *The Political Economy of Robots: Prospects for Prosperity and Peace in the Automated 21st Century*, ed. Ryan Kiggins (Cham: Springer International Publishing, 2018), 3–4.

7. *Summary of the 2018 National Defense Strategy of the United States of America* (US Department of Defense, January 2018), 3, https://dod.defense.gov/Portals/1/Documents/pubs/2018-National-Defense-Strategy-Summary.pdf.

8. For an overview of how several nations are approaching AI in war, see Forrest E. Morgan et al., *Military Applications of Artificial Intelligence* (RAND Corporation, 2020), https://www.rand.org/content/dam/rand/pubs/research_reports/RR3100/RR3139-1/RAND_RR3139-1.pdf.

9. Sydney J. Freedberg Jr., "AI Can Save Taxpayers Money: JAIC Chief," *Breaking Defense*, January 14, 2021, https://breakingdefense.com/2021/01/ai-can-save-taxpayers-money-jaic-chief/.

10. Melania Scerra, "Spending Forecast—Military Robotics Globally 2025," *Statista*, April 16, 2020, https://www.statista.com/statistics/441960/forecast-for-military-robotic-market-spending-worldwide/; Justin Haner and Denise Garcia, "The Artificial Intelligence Arms Race: Trends and World Leaders in Autonomous Weapons Development," *Global Policy* 10, no. 3 (2019): 331.

11. "Open Letter on Autonomous Weapons," Future of Life Institute, July 28, 2015, https://futureoflife.org/open-letter-autonomous-weapons/; "About ICRAC," International Committee for Robot Arms Control, February 9, 2012, https://www.icrac.net/about-icrac/.

12. See, for example, Michael C. Horowitz, "The Ethics and Morality of Robotic Warfare: Assessing the Debate over Autonomous Weapons," *Daedalus* 145, no. 4 (Fall 2016): 26, https://direct.mit.edu/daed/article/145/4/25/27111/The-Ethics-amp-Morality-of-Robotic-Warfare.

13. Kelsey D. Atherton, "Are Killer Robots the Future of War? Parsing the Facts on Autonomous Weapons," *New York Times*, November 15, 2018, https://www.nytimes.com/2018/11/15/magazine/autonomous-robots-weapons.html; see also Christian Brose, *The Kill Chain: Defending America in the Future of High-Tech Warfare* (New York:

Hachette Books, 2020); Larry Lewis, *Redefining Human Control: Lessons from the Battlefield* (Center for Naval Analyses, March 2018), https://www.cna.org/cna_files/pdf/DOP-2018-U-017258-Final.pdf.

14. Interview with the authors, April 28, 2020.

15. Maja Zehfuss, "Targeting: Precision and the Production of Ethics," *European Journal of International Relations* 17, no. 3 (September 1, 2011): 550.

16. William A. Owens, "The Emerging U.S. System-of-Systems," Strategic Forum, February 1996, https://web.archive.org/web/20100105160638/http://www.ndu.edu/inss/strforum/SF_63/forum63.html; *Joint Vision 2010* (Washington, DC: Chairman of the Joint Chiefs of Staff, July 2010), https://web.archive.org/web/20161224220150/http://www.dtic.mil/jv2010/jv2010.pdf.

17. John T. Correll, "The Emergence of Smart Bombs," *Air Force Magazine*, March 2010, 64, https://www.airforcemag.com/PDF/MagazineArchive/Documents/2010/March%202010/0310bombs.pdf.

18. M. Taylor Fravel, *Active Defense: China's Military Strategy Since 1949* (Princeton, NJ: Princeton University Press, 2019), 7.

19. Interview with the authors.

20. Sharon Weinberger, "The Return of the Pentagon's Yoda," *Foreign Policy*, September 12, 2018, https://foreignpolicy.com/2018/09/12/the-return-of-the-pentagons-yoda-andrew-marshall/.

21. *Annual Report to Congress: Military and Security Developments Involving the People's Republic of China 2019* (US Department of Defense, May 2019), https://media.defense.gov/2019/May/02/2002127082/-1/-1/1/2019_CHINA_MILITARY_POWER_REPORT.pdf.

22. Kania, "'AI Weapons' in China's Military Innovation," 2–4.

23. Brose, *The Kill Chain*, 111–112; "An Interview with Robert O. Work," *Joint Forces Quarterly* 84, no. 1 (2017), https://ufdcimages.uflib.ufl.edu/AA/00/06/15/87/00084/1st%20Quarter-2017.pdf.

24. Interview with the authors; Cheryl Pellerin, "Deputy Secretary: Third Offset Strategy Bolsters America's Military Deterrence," Department of Defense, October 31, 2016, https://www.defense.gov/Explore/News/Article/Article/991434/deputy-secretary-third-offset-strategy-bolsters-americas-military-deterrence/; Paul McLeary, "The Pentagon's Third Offset May Be Dead, But No One Knows What Comes Next," *Foreign Policy*, December 18, 2017, https://foreignpolicy.com/2017/12/18/the-pentagons-third-offset-may-be-dead-but-no-one-knows-what-comes-next/; "Report to Congress on Emerging Military Technologies," United States Naval Institute, July 21, 2020, https://news.usni.org/2020/07/21/report-to-congress-on-emerging-military-technologies.

25. *DoD Directive 3000.09: Autonomy in Weapons Systems* (US Department of Defense, November 21, 2012), https://www.esd.whs.mil/Portals/54/Documents/DD/issuances/dodd/300009p.pdf.

26. Interview with the authors.

27. Interview with the authors. For a holistic overview of the capabilities the Pentagon is pursuing, see Margarita Konaev et al., *U.S. Military Investments in Autonomy and AI: A Strategic Assessment* (Georgetown University Center for Security and Emerging Technology, October 2020), https://cset.georgetown.edu/wp-content/uploads/CSET-U.S.-Military-Investments-in-Autonomy-and-AI-A-Budgetary-Assessment.pdf.

28. Interview with the authors; Ryan Pickrell, "The Pentagon Is Reviving This Hard-Hitting Cold War Strategy to Devastate Any Chinese and Russian Assaults," *Business Insider*, March 5, 2019, https://www.businessinsider.com/dod-revives-cold-war-strategy-to-crush-russian-and-chinese-assaults-2019-3; Justin Doubleday, "DARPA Advancing 'Assault Breaker II' to Test Technologies Underpinning Multi-Domain Operations," *Inside Defense*, May 2, 2019, https://insidedefense.com/daily-news/darpa-advancing-assault-breaker-ii-test-technologies-underpinning-multi-domain-operations; *Study on Countering Anti-access Systems with Longer Range and Standoff Capabilities: Assault Breaker II* (Washington, DC: Defense Science Board, June 2018), https://dsb.cto.mil/reports/2010s/LRE%20Executive%20Summary__Final.pdf.

29. Paul Scharre, *Army of None: Autonomous Weapons and the Future of War* (New York: W. W. Norton, 2018), 97.

30. "Establishing the CODE for Unmanned Aircraft to Fly as Collaborative Teams," DARPA, January 21, 2015, https://www.darpa.mil/news-events/2015-01-21.

31. "In the Sky and on the Ground, Collaboration Vital to DARPA's CODE for Success," DARPA, March 22, 2019, https://www.darpa.mil/news-events/2019-03-22.

32. DARPA, "Our CODE technology for collaborative, autonomous systems is set to transition to @USNavy! Naval Air Systems Command (@NAVAIRnews) will take ownership of CODE after DARPA closes out its role in the program this year. https://go.usa.gov/xEJve #autonomy," Twitter, March 28, 2019, 1:04 p.m., https://twitter.com/darpa/status/1110995915010129927.

33. Despite the similar naming conventions, this system has no relationship to DeepMind.

34. Patrick Tucker, "An AI Just Beat a Human F-16 Pilot In a Dogfight—Again," *DefenseOne*, August 20, 2020, https://www.defenseone.com/technology/2020/08/ai-just-beat-human-f-16-pilot-dogfight-again/167872/. For more on AI in air operations, see Aaron Gregg, "In a First, Air Force Uses AI on Military Jet," *Washington Post*, December 16, 2020, https://www.washingtonpost.com/business/2020/12/16/air-force-artificial-intelligence/.

35. Nathan Strout, "Inside the Army's Futuristic Test of Its Battlefield Artificial Intelligence in the Desert," *C4ISRNet*, September 25, 2020, https://www.c4isrnet.com/artificial-intelligence/2020/09/25/the-army-just-conducted-a-massive-test-of-its-battlefield-artificial-intelligence-in-the-desert/.

36. Sydney J. Freedberg Jr., "Let Your Robots Off the Leash—Or Lose: AI Experts," *Breaking Defense*, November 19, 2020, https://breakingdefense.com/2020/11/let-your-robots-off-the-leash-or-lose-ai-experts/.

37. Will Knight, "The Pentagon Inches Toward Letting AI Control Weapons," *Wired*, May 10, 2021, https://www.wired.com/story/pentagon-inches-toward-letting -ai-control-weapons/.

38. Horowitz, "The Ethics and Morality of Robotic Warfare," 25–36.

39. Strout, "Inside the Army's Futuristic Test of Its Battlefield Artificial Intelligence in the Desert."

40. The current version of the Sea Hunter does not carry weapons. Scharre, *Army of None*, 79.

41. Interview with the authors, May 5, 2020.

42. The group dissolved in 2013.

43. Interview with the authors; William Burr, ed., "Launch on Warning: The Development of U.S. Capabilities, 1959–1979," The National Security Archive, April 2001, https://nsarchive2.gwu.edu/NSAEBB/NSAEBB43/.

44. "Member: Lucy Suchman," International Committee for Robot Arms Control, accessed July 1, 2020, https://www.icrac.net/members/lucy-suchman/.

45. Interview with the authors.

46. Lucy Suchman, "Situational Awareness and Adherence to the Principle of Distinction as a Necessary Condition for Lawful Autonomy," in *Lethal Autonomous Weapons Systems: Technology, Definition, Ethics, Law & Security*, ed. Robin Geiß (Berlin: German Federal Foreign Office, 2017), 4.

47. Interview with the authors.

48. Interview with the authors. Others have raised similar concerns. Arthur Holland Michel, "Known Unknowns: Data Issues and Military Autonomous Systems," United Nations Institute for Disarmament Research, May 2021, https://www.unidir.org /sites/default/files/2021-05/Holland_KnownUnknowns_20210517_0.pdf.

49. *Formal Investigation into the Circumstances Surrounding the Downing of Iran Air Flight 655 on 3 July 1988* (United States Navy, August 19, 1988), 13, https://www.jag .navy.mil/library/investigations/VINCENNES%20INV.pdf; Richard Danzig, *Technology Roulette* (Washington, DC: Center for a New American Security, June 2018), 14, https://s3.us-east-1.amazonaws.com/files.cnas.org/documents/CNASReport-Technology -Roulette-DoSproof2v2.pdf.

50. Interview with the authors.

51. Scharre, *Army of None*, 137–144.

52. Interview with the authors.

53. Sydney J. Freedberg Jr., "How AI Could Change the Art of War," *Breaking Defense*, April 25, 2019, https://breakingdefense.com/2019/04/how-ai-could-change -the-art-of-war/. For more on trust in weapons systems, see Heather M. Roff and David Danks, "'Trust But Verify': The Difficulty of Trusting Autonomous Weapons Systems," *Journal of Military Ethics* 17, no. 1 (2018): 2–20.

54. Rupert Ticehurst, "The Martens Clause and the Laws of Armed Conflict," International Committee of the Red Cross, April 30, 1997, https://www.icrc.org/en

/doc/resources/documents/article/other/57jnhy.htm; "Heed the Call," Human Rights Watch, August 21, 2018, https://www.hrw.org/report/2018/08/21/heed-call/moral -and-legal-imperative-ban-killer-robots; Scharre, *Army of None*, 263–266.

55. Neil Davison, "A Legal Perspective: Autonomous Weapons under International Humanitarian Law," in *Perspectives on Lethal Autonomous Weapon Systems*, UNODA Occasional Papers (New York: United Nations Publications, 2018), 7–8.

56. Interview with the authors.

57. Interview with the authors.

58. "Country Views on Killer Robots," Campaign to Stop Killer Robots, March 11, 2020, https://www.stopkillerrobots.org/wp-content/uploads/2020/03/KRC_Country Views_11Mar2020.pdf.

59. "Open Letter on Autonomous Weapons."

60. Samuel Gibbs, "Elon Musk Leads 116 Experts Calling for Outright Ban of Killer Robots," *Guardian*, August 20, 2017, http://www.theguardian.com/technology/2017 /aug/20/elon-musk-killer-robots-experts-outright-ban-lethal-autonomous-weapons -war; Arjun Kharpal, "Stephen Hawking Says A.I. Could Be 'Worst Event in the History of Our Civilization,'" *CNBC*, November 6, 2017, https://www.cnbc.com/2017/11/06 /stephen-hawking-ai-could-be-worst-event-in-civilization.html; "Open Letter on Autonomous Weapons"; Elon Musk, "It begins . . . ," Twitter, September 4, 2017, 2:11 a.m., https://twitter.com/elonmusk/status/904638455761612800?s=20.

61. "Secretary-General's Address to the Paris Peace Forum," United Nations Secretary-General, November 11, 2018, https://www.un.org/sg/en/content/sg/statement /2018-11-11/allocution-du-secr%C3%A9taire-g%C3%A9n%C3%A9ral-au-forum-de -paris-sur-la-paix; Patrick Tucker, "Red Cross Calls for More Limits on Autonomous Weapons," *Defense One*, May 13, 2021, https://www.defenseone.com/technology /2021/05/red-cross-calls-limits-autonomous-weapons/174018/. For an excellent over-view of the subject, see Hayley Evans and Natalie Salmanowitz, "Lethal Autonomous Weapons Systems: Recent Developments," *Lawfare*, March 7, 2019, https://www .lawfareblog.com/lethal-autonomous-weapons-systems-recent-developments. It's worth noting that public opinion on this subject may be contingent on the context. As the political scientist Michael Horowitz finds, the US public is more willing to permit the development of autonomous weapons to protect American military forces. Michael C. Horowitz, "Public Opinion and the Politics of the Killer Robots Debate," *Research & Politics* 3, no. 1 (2016), https://journals.sagepub.com/doi/pdf/10.1177 /2053168015627183.

62. This expression is from the 1983 movie *WarGames*.

63. Interview with the authors. Robert Work, "Principles for the Combat Employ-ment of Weapon Systems with Autonomous Functionalities," Center for a New Amer-ican Security, April 28, 2021, https://www.cnas.org/publications/reports/proposed -dod-principles-for-the-combat-employment-of-weapon-systems-with-autonomous -functionalities, 9.

64. "Position Paper Submitted by the Chinese Delegation," Certain Conventional Weapons Review Conference, April 9, 2018, https://web.archive.org/web/2020

1101064523/https://www.unog.ch/80256EDD006B8954/(httpAssets)/E42AE83BDB3 525D0C125826C0040B262/$file/CCW_GGE.1_2018_WP.7.pdf.

65. "Position Paper Submitted by the Chinese Delegation."

66. Elsa Kania, "China's Strategic Ambiguity and Shifting Approach to Lethal Autonomous Weapons Systems," *Lawfare*, April 17, 2018, https://www.lawfareblog .com/chinas-strategic-ambiguity-and-shifting-approach-lethal-autonomous-weapons -systems; Chan, "Could China Develop Killer Robots in the Near Future?"

67. Patrick Tucker, "SecDef: China Is Exporting Killer Robots to the Mideast," *DefenseOne*, November 5, 2019, https://www.defenseone.com/technology/2019/11 /secdef-china-exporting-killer-robots-mideast/161100/; Michael Horowitz et al., "Who's Prone to Drone? A Global Time-series Analysis of Armed Uninhabited Aerial Vehicle Proliferation," *Conflict Management and Peace Science* (2020).

68. Scharre, *Army of None*, 82. See also Robert O. Work, "A Joint Warfighting Concept for Systems Warfare," Center for a New American Security, December 17, 2020, https://www.cnas.org/publications/commentary/a-joint-warfighting-concept-for -systems-warfare.

69. Interview with the authors.

70. Interview with the authors.

71. Jeffrey Dastin and Paresh Dave, "US Has 'Moral Imperative' to Develop AI Weapons, Says Panel," *Reuters*, January 26, 2021, https://www.reuters.com/article/us -usa-military-ai/u-s-commission-cites-moral-imperative-to-explore-ai-weapons-id USKBN29V2M0.

72. As discussed in chapter 5, the Department of Defense has laid out its AI principles.

73. Defense Science Board, *Summer Study on Autonomy* (US Department of Defense, June 2016), 16, https://www.hsdl.org/?view&did=794641.

74. Scharre, *Army of None*, 95.

75. Paul Scharre, "A Million Mistakes a Second," *Foreign Policy*, September 12, 2018, https://foreignpolicy.com/2018/09/12/a-million-mistakes-a-second-future-of-war/.

76. Scharre, *Army of None*, 95.

77. Interview with the authors.

78. Interview with the authors.

79. Scharre, *Army of None*, 101.

80. Scharre, *Army of None*, 101.

81. Note that some democracies, such as Israel, have been more willing than others to build lethal autonomous weapons. The Israeli Harpy is one of the world's most notable examples. This trend suggests that the threat environment may be as important as regime type when it comes to new weapons development.

82. *Law of War Manual* (Office of General Counsel, Department of Defense, December 2016), 64, https://dod.defense.gov/Portals/1/Documents/pubs/DoD%20Law%20

of%20War%20Manual%20-%20June%202015%20Updated%20Dec%202016
.pdf?ver=2016-12-13-172036-190.

83. For a discussion of their evolution, see Charlie Savage, *Power Wars: The Relentless Rise of Presidential Authority and Secrecy* (New York: Little, Brown, 2017).

84. Sydney J. Freedberg Jr., "'We May Be Losing the Race' for AI with China: Bob Work," *Breaking Defense*, September 2, 2020, https://breakingdefense.com/2020/09/we-may-be-losing-the-race-for-ai-with-china-bob-work/.

85. Work and Scharre, "Transcript from Military AI Applications."

86. Sydney J. Freedberg Jr., "JAIC Chief Asks: Can AI Prevent Another 1914?," *Breaking Defense*, November 11, 2020, https://breakingdefense.com/2020/11/jaic-chief-asks-can-ai-prevent-another-1914/.

87. Interview with the authors.

CHAPTER 7

1. Brian Fung, "The NSA Hacks Other Countries by Buying Millions of Dollars' Worth of Computer Vulnerabilities," *Washington Post*, August 31, 2013, https://www.washingtonpost.com/news/the-switch/wp/2013/08/31/the-nsa-hacks-other-countries-by-buying-millions-of-dollars-worth-of-computer-vulnerabilities/; "Vulnerabilities Equities Policy and Process for the United States Government," The White House, November 15, 2017, https://web.archive.org/web/20200329180821/https://www.whitehouse.gov/sites/whitehouse.gov/files/images/External%20-%20Unclassified%20VEP%20Charter%20FINAL.PDF; Ben Buchanan, "The Life Cycles of Cyber Threats," *Survival* 58, no. 1 (2016), http://www.iiss.org/en/publications/survival/sections/2016-5e13/survival--global-politics-and-strategy-february-march-2016-44d5/58-1-03-buchanan-7bfc; Jason Healey, "The Cyber Budget Shows What the U.S. Values—and It Isn't Defense," *Lawfare*, June 1, 2020, https://www.lawfareblog.com/cyber-budget-shows-what-us-values%E2%80%94and-it-isnt-defense.

2. For more on hacking and geopolitics, see Ben Buchanan, *The Hacker and the State: Cyber Attacks and the New Normal of Geopolitics* (Cambridge, MA: Harvard University Press, 2020).

3. Joshua Davis, "Say Hello to Stanley," *Wired*, January 1, 2006, https://www.wired.com/2006/01/stanley/.

4. Interview with the authors, August 25, 2020.

5. Cade Metz, "DARPA Goes Full Tron with Its Grand Battle of the Hack Bots," *Wired*, July 5, 2016, https://www.wired.com/2016/07/-trashed-19/.

6. Metz, "DARPA Goes Full Tron with Its Grand Battle of the Hack Bots"; *60 Minutes*, "DARPA's Cyber Grand Challenge," YouTube video, February 8, 2015, 00:48, https://www.youtube.com/watch?v=OVV_k73z3E0.

7. Thanassis Avgerinos et al., "The Mayhem Cyber Reasoning System," *IEEE Security Privacy* 16, no. 2 (2018): 52–54.

8. Brumley had also used the name Mayhem for previous autonomous vulnerability discovery systems.

9. Interview with the authors, September 14, 2020.

10. National Academies of Sciences, Engineering, and Medicine et al., *Implications of Artificial Intelligence for Cybersecurity: Proceedings of a Workshop* (Washington, D.C.: National Academies Press, 2020), 12. Since then, machine learning vulnerability discovery has improved, though the relative value of machine learning in software vulnerability discovery remains a subject of debate. Konstantin Böttinger, Patrice Godefroid, and Rishabh Singh, "Deep Reinforcement Fuzzing," in *2018 IEEE Security and Privacy Workshops* (San Francisco, 2018), 116–122; Augustus Odena et al., "Tensorfuzz: Debugging Neural Networks with Coverage-Guided Fuzzing," in *Proceedings of the 36th International Conference on Machine Learning* 97 (2019): 4901–4911; Patrice Godefroid, Hila Peleg, and Rishabh Singh, "Learn&fuzz: Machine Learning for Input Fuzzing," in *2017 32nd IEEE/ACM International Conference on Automated Software Engineering* (Urbana-Champaign, IL: IEEE Press, 2017), 50–59; Ari Takanen et al., *Fuzzing for Software Security Testing and Quality Assurance*, 2nd ed. (Boston: Artech House, 2018); Gary J. Saavedra et al., "A Review of Machine Learning Applications in Fuzzing," arXiv, June 13, 2019, http://arxiv.org/abs/1906.11133.

11. Interview with the authors.

12. Interview with the authors; Avgerinos et al., "The Mayhem Cyber Reasoning System," 54–58.

13. National Academies of Sciences, Engineering, and Medicine et al., *Implications of Artificial Intelligence for Cybersecurity*, 12.

14. Robert Vamosi, "The Hacker Mind Podcast: Inside DARPA's Cyber Grand Challenge," ForAllSecure, August 21, 2020, https://forallsecure.com/blog/the-hacker-mind-podcast-inside-darpas-cyber-grand-challenge; interview with the authors.

15. Interview with the authors.

16. "DARPA Celebrates Cyber Grand Challenge Winners," DARPA, August 5, 2016, https://www.darpa.mil/news-events/2016-08-05a.

17. Cade Metz, "Hackers Don't Have to Be Human Anymore. This Bot Battle Proves It," *Wired*, August 5, 2016, https://www.wired.com/2016/08/security-bots-show-hacking-isnt-just-humans/.

18. Interview with the authors; "DARPA Celebrates Cyber Grand Challenge Winners."

19. Interview with the authors.

20. Robert Vamosi, "The Hacker Mind Podcast: Can a Machine Think Like a Hacker?," ForAllSecure, September 2, 2020, https://blog.forallsecure.com/the-hacker-mind-podcast-can-a-machine-think-like-a-hacker.

21. Vamosi, "The Hacker Mind Podcast: Can a Machine Think Like a Hacker?"

22. Interviews with the authors.

23. Tom Simonite, "This Bot Hunts Software Bugs for the Pentagon," *Wired*, June 1, 2020, https://www.wired.com/story/bot-hunts-software-bugs-pentagon/.

24. For insights on automating this process and its hallmarks, see Thomas Rid and Ben Buchanan, "Attributing Cyber Attacks," *Journal of Strategic Studies* 39, no. 1 (2015): 4–37.

25. Michael V. Hayden, "The Making of America's Cyberweapons," *Christian Science Monitor*, February 24, 2016, https://www.csmonitor.com/World/Passcode /Passcode-Voices/2016/0224/The-making-of-America-s-cyberweapons.

26. Zachary Fryer-Biggs, "Secretive Pentagon Research Program Looks to Replace Human Hackers with AI," *Yahoo News*, September 13, 2020, https://news.yahoo .com/secretive-pentagon-research-program-looks-to-replace-human-hackers-with -ai-090032920.html. During one major operation, US operators manually crossed off targets on a sheet of paper hung on a wall. Garrett Graff, "The Man Who Speaks Softly—and Commands a Big Cyber Army," *Wired*, October 13, 2020, https://www .wired.com/story/general-paul-nakasone-cyber-command-nsa/.

27. Fryer-Biggs, "Secretive Pentagon Research Program Looks to Replace Human Hackers with AI."

28. Noah Shachtman, "This Pentagon Project Makes Cyberwar as Easy as Angry Birds," *Wired*, May 28, 2013, https://web.archive.org/web/20160423111428/http:// www.wired.com/2013/05/pentagon-cyberwar-angry-birds/all/.

29. Fryer-Biggs, "Secretive Pentagon Research Program Looks to Replace Human Hackers with AI."

30. Interview with the authors, December 3, 2020.

31. Fryer-Biggs, "Secretive Pentagon Research Program Looks to Replace Human Hackers with AI."

32. Fryer-Biggs, "Secretive Pentagon Research Program Looks to Replace Human Hackers with AI."

33. Fryer-Biggs, "Secretive Pentagon Research Program Looks to Replace Human Hackers with AI."

34. Fryer-Biggs, "Secretive Pentagon Research Program Looks to Replace Human Hackers with AI."

35. For a book-length history of the operation, see Kim Zetter, *Countdown to Zero Day* (New York: Crown, 2014).

36. Fryer-Biggs, "Secretive Pentagon Research Program Looks to Replace Human Hackers with AI"; interview with the authors.

37. Paul M. Nakasone and Michael Sulmeyer, "How to Compete in Cyberspace," *Foreign Affairs*, August 25, 2020; Ellen Nakashima, "U.S. Cyber Command Operation Disrupted Internet Access of Russian Troll Factory on Day of 2018 Midterms," *Washington Post*, February 27, 2019, https://www.washingtonpost.com/world/national -security/us-cyber-command-operation-disrupted-internet-access-of-russian-troll

-factory-on-day-of-2018-midterms/2019/02/26/1827fc9e-36d6-11e9-af5b-b51b7
ff322e9_story.html.

38. Patrick Tucker, "Trump's Pick for NSA/CyberCom Chief Wants to Enlist AI For Cyber Offense," *DefenseOne*, January 9, 2018, https://www.defenseone.com /technology/2018/01/how-likely-next-nsacybercom-chief-wants-enlist-ai/145085/.

39. Nakasone and Sulmeyer, "How to Compete in Cyberspace."

40. Andy Greenberg, "The Untold Story of NotPetya, the Most Devastating Cyber-attack in History," *Wired*, August 22, 2018, https://www.wired.com/story/notpetya -cyberattack-ukraine-russia-code-crashed-the-world/; Scott J. Shackelford et al., "From Russia with Love: Understanding the Russian Cyber Threat to US Critical Infrastructure and What to Do About It," *Nebraska Law Review* 96 (2017): 320; Ben Buchanan and Michael Sulmeyer, *Russia and Cyber Operations: Challenges and Opportunities for the next US Administration* (Carnegie Endowment for International Peace, December 16, 2016), https://carnegieendowment.org/files/12-16-16_Russia_and _Cyber_Operations.pdf.

41. Ondrej Kubovič, Juraj Jánošík, and Peter Košinár, *Machine-Learning Era in Cybersecurity: A Step Towards a Safer World or the Brink of Chaos?* (ESET, February 2019), 11, https://www.eset.com/fileadmin/ESET/US/download/ESETus-Machine-Learning-Era -in-Cybersecurity-Whitepaper-WEB.pdf.

42. Interview with the authors.

43. Interview with the authors; Ben Buchanan, *The Legend of Sophistication in Cyber Operations* (Harvard University Belfer Center for Science and International Affairs, January 2017), https://www.belfercenter.org/sites/default/files/files/publication /Legend%20Sophistication%20-%20web.pdf.

44. Interview with the authors, September 9, 2020.

45. Thomas Rid, *Rise of the Machines* (New York: W. W. Norton, 2016), 328.

46. Interview with the authors.

47. *Evolution of Malware Prevention* (Microsoft, n.d.), 4, https://info.microsoft.com /rs/157-GQE-382/images/Windows%20Defender%20ML%20Whitepaper.pdf.

48. This is in part due to the rise of ransomware, which by definition makes itself known to the target. *M-Trends 2021* (FireEye, April 13, 2021), 11, https://content .fireeye.com/m-trends/rpt-m-trends-2021.

49. *Venture Investment at the Nexus of Cybersecurity and AI/ML* (Dell Technologies Capital, June 5, 2019), 2, https://files.pitchbook.com/website/files/pdf/Dell _Technologies_Capital_Investment_at_Nexus_of_Cybersecurity_AI_ML_fJk.pdf.

50. Richard Lippmann et al., "The 1999 DARPA Off-Line Intrusion Detection Evaluation," *Computer Networks* 34, no. 4 (2000): 579–595.

51. Stephen Verbeke, "COVID-19's Impact on Cybersecurity Incident Response," Novetta, May 26, 2020, https://www.novetta.com/2020/05/cyber-covid/. This has upended more than just cybersecurity algorithms. Will Douglas Heaven, "Our Weird Behavior during the Pandemic Is Messing with AI Models," *MIT Technology Review*,

May 11, 2020, https://www.technologyreview.com/2020/05/11/1001563/covid
-pandemic-broken-ai-machine-learning-amazon-retail-fraud-humans-in-the-loop/.

52. *Evolution of Malware Prevention*, 4.

53. Hemant Rathore et al., "Malware Detection Using Machine Learning and Deep Learning," arXiv, April 4, 2019, http://arxiv.org/abs/1904.02441.

54. Another pioneer, Yann LeCun, is often included in this group. The three won the Turing Award, the highest honor in computer science, in 2018.

55. Ian Goodfellow, "Defense Against the Dark Arts: An Overview of Adversarial Example Security Research and Future Research Directions," arXiv, June 11, 2018, 6, http://arxiv.org/abs/1806.04169.

56. Goodfellow, "Defense Against the Dark Arts," 8.

57. This illusion also extends beyond humans to other creatures. Kazuo Fujita, Donald S. Blough, and Patricia M. Blough, "Pigeons See the Ponzo Illusion," *Animal Learning & Behavior* 19, no. 3 (1991): 283–293; Kathryn A. L. Bayne and Roger T. Davis, "Susceptibility of Rhesus Monkeys (Macaca Mulatta) to the Ponzo Illusion," *Bulletin of the Psychonomic Society* 21, no. 6 (1983): 476–478. For another example, see Stephen Law, "Do You See a Duck or a Rabbit: Just What Is Aspect Perception?," *Aeon*, July 31, 2018, https://aeon.co/ideas/do-you-see-a-duck-or-a-rabbit-just-what-is-aspect-perception.

58. For example, see Ian J. Goodfellow, Jonathon Shlens, and Christian Szegedy, "Explaining and Harnessing Adversarial Examples," arXiv, December 20, 2014, https://arxiv.org/abs/1412.6572; Kevin Eykholt et al., "Robust Physical-World Attacks on Deep Learning Visual Classification," in *Proceedings of the IEEE Conference on Computer Vision and Pattern Recognition* (Salt Lake City: 2018), 1625–1634.

59. Goodfellow, "Defense Against the Dark Arts," 11–14.

60. Mahmood Sharif et al., "Accessorize to a Crime: Real and Stealthy Attacks on State-of-the-Art Face Recognition," in *Proceedings of the 2016 ACM SIGSAC Conference on Computer and Communications Security* (New York: Association for Computing Machinery, 2016), 1528–1540; Eykholt et al., "Robust Physical-World Attacks on Deep Learning Visual Classification"; Anish Athalye et al., "Fooling Neural Networks in the Physical World with 3D Adversarial Objects," LabSix, October 31, 2017, https://www.labsix.org/physical-objects-that-fool-neural-nets/.

61. Karen Hao, "A New Set of Images That Fool AI Could Help Make It More Hacker-Proof," *MIT Technology Review*, June 21, 2019, https://www.technologyreview.com/2019/06/21/828/a-new-set-of-images-that-fool-ai-could-help-make-it-more-hacker-proof/.

62. Nicholas Carlini, "A Complete List of All (ArXiv) Adversarial Example Papers," June 15, 2019, https://nicholas.carlini.com/writing/2019/all-adversarial-example-papers.html; Alexey Kurakin et al., "Adversarial Attacks and Defences Competition," arXiv, March 31, 2018, http://arxiv.org/abs/1804.00097.

63. Ian Goodfellow and Nicolas Papernot, "Is Attacking Machine Learning Easier than Defending It?," *Cleverhans*, February 15, 2017, http://www.cleverhans.io/security

/privacy/ml/2017/02/15/why-attacking-machine-learning-is-easier-than-defending
-it.html.

64. Cade Metz, "How to Fool AI into Seeing Something That Isn't There," *Wired*, July 29, 2016, https://www.wired.com/2016/07/fool-ai-seeing-something-isnt/.

65. Garrett Reim, "US Air Force Grapples with Vexing Problem of AI Spoofing," FlightGlobal, September 1, 2020, https://www.flightglobal.com/defence/us-air-force -grapples-with-vexing-problem-of-ai-spoofing/139973.article.

66. Matt Fredrikson, Somesh Jha, and Thomas Ristenpart, "Model Inversion Attacks That Exploit Confidence Information and Basic Countermeasures," in *Proceedings of the 22nd ACM SIGSAC Conference on Computer and Communications Security* (New York: Association for Computing Machinery, 2015), 1322–1333; Reza Shokri et al., "Membership Inference Attacks Against Machine Learning Models," in *2017 IEEE Symposium on Security and Privacy*, https://ieeexplore.ieee.org/document/7958568, 3–18; Florian Tramèr et al., "Stealing Machine Learning Models via Prediction APIs," in *Proceedings of the 25th USENIX Security Symposium* (Austin, TX: USENIX Association, 2016), 601–618; Yuheng Zhang et al., "The Secret Revealer: Generative Model-Inversion Attacks Against Deep Neural Networks," arXiv, November 17, 2019, http://arxiv.org/abs/1911.07135; Nicholas Carlini et al., "Extracting Training Data from Large Language Models," arXiv, December 14, 2020, http://arxiv.org/abs /2012.07805.

67. For an excellent overview of machine learning's weaknesses, see Andrew J. Lohn, *Hacking AI* (Georgetown University Center for Security and Emerging Technology, December 2020), https://cset.georgetown.edu/wp-content/uploads/CSET -Hacking-AI.pdf.

68. Goodfellow and Papernot, "Is Attacking Machine Learning Easier than Defending It?"

69. Irwin Lebow, "The Impact of Computer Security in the Department of Defense," in *Second Seminar on the DOD Computer Security Initiative Program* (Gaithersburg, MD: National Bureau of Standards, 1980), C-19, https://csrc.nist.gov/csrc/media /publications/conference-paper/1980/01/15/proceedings-2nd-seminar-dod-computer -security-initiative/documents/1980-2nd-seminar-proceedings.pdf.

70. Bruce Schneier, "Attacking Machine Learning Systems," *Computer* 53, no. 5 (2020): 78–80.

71. Nicholas Carlini, "On Evaluating Adversarial Robustness" (video), 2019 Conference on Applied Machine Learning in Information Security, October 26, 2019, 32:10, https://www.camlis.org/2019/keynotes/carlini.

72. "Defending Against Adversarial Artificial Intelligence," DARPA, February 6, 2019, https://www.darpa.mil/news-events/2019-02-06.

73. The intersection of AI and cybersecurity could affect escalation risks as well. Wyatt Hoffman, *AI and the Future of Cyber Competition* (Georgetown University Center for Security and Emerging Technology, January 2021), https://cset .georgetown.edu/wp-content/uploads/CSET-AI-and-the-Future-of-Cyber-Competition

.pdf; Ben Buchanan, *The Cybersecurity Dilemma: Hacking, Trust, and Fear between Nations* (New York: Oxford University Press, 2016).

CHAPTER 8

1. Thomas Rid, *Active Measures: The Secret History of Disinformation and Political Warfare* (New York: Farrar, Straus and Giroux, 2020), 302–303.

2. *Soviet Influence Activities: A Report on Active Measures and Propaganda, 1986–87* (US Department of State, August 1987), 3, 33–51, https://www.globalsecurity.org/intell /library/reports/1987/soviet-influence-activities-1987.pdf.

3. Rid, *Active Measures: The Secret History of Disinformation and Political Warfare*, 303.

4. Rid, *Active Measures*, 306.

5. Rid, *Active Measures*, 306–310; Lola Ogunnaike, "The Passion of Kanye West," *Rolling Stone*, February 9, 2006, https://www.rollingstone.com/music/music-news /the-passion-of-kanye-west-71551/.

6. Philip Ewing, "Russia's Election Meddling Part of a Long History of 'Active Measures,'" *NPR*, May 23, 2017, https://www.npr.org/2017/05/23/528500501/lies-forgery -and-skulduggery-the-long-history-of-active-measures.

7. Rid, *Active Measures*, 4.

8. Lily Hay Newman, "The Election Threats That Keep US Intelligence Up at Night," *Wired*, September 28, 2020, https://www.wired.com/story/election-threats-fbi-cisa -disinformation-hack-leak/; Thomas Rid and Ben Buchanan, "Hacking Democracy," *SAIS Review of International Affairs* 38, no. 1 (2018): 3–16; Scott J. Shackelford et al., "Making Democracy Harder to Hack," *University of Michigan Journal of Law Reform* 50 (2016): 629. On the interplay between foreign and domestic sources of disinformation, see Katerina Sedova, *The AI Future of Influence Campaigns* (Georgetown University Center for Security and Emerging Technology, 2021).

9. Interviews with the authors, December 16, 2019 and February 25, 2020.

10. Interview with the authors. See also Rand Waltzman, "The Story Behind the DARPA Social Media in Strategic Communication (SMISC) Program," Information Professionals Association, June 28, 2017, https://information-professionals.org /the-darpa-social-media-in-strategic-communication-smisc-program/.

11. "Form 10-Q," United States Securities and Exchange Commission, August 11, 2014, https://www.sec.gov/Archives/edgar/data/1418091/000156459014003474/twtr -10q_20140630.htm. The study examined Twitter links to the top 2,135 most popular sites on the internet. Stefan Wojcik et al., "Twitter Bots: An Analysis of the Links Automated Accounts Share," Pew Research Center, April 9, 2018, https://www .pewresearch.org/internet/2018/04/09/bots-in-the-twittersphere/.

12. Paris Martineau, "What Is a Bot?" *Wired*, November 16, 2018, https://www.wired .com/story/the-know-it-alls-what-is-a-bot/.

13. V. S. Subrahmanian et al., "The DARPA Twitter Bot Challenge," *Computer* 49, no. 6 (2016): 1.

14. Subrahmanian et al., "The DARPA Twitter Bot Challenge," 2.

15. Subrahmanian et al., "The DARPA Twitter Bot Challenge," 5–11.

16. Subrahmanian et al., "The DARPA Twitter Bot Challenge," 11.

17. Subrahmanian et al., "The DARPA Twitter Bot Challenge," 10–11.

18. Interview with the authors.

19. For a discussion of this, see Seva Gunitsky, "Democracies Can't Blame Putin for Their Disinformation Problem," *Foreign Policy*, April 21, 2020, https://foreignpolicy .com/2020/04/21/democracies-disinformation-russia-china-homegrown/.

20. Brad Stone, *The Everything Store: Jeff Bezos and the Age of Amazon* (New York: Random House, 2013), 133.

21. Stone, *The Everything Store*, 132–134.

22. Tom Simonite, "The People Trying to Make Internet Recommendations Less Toxic," *Wired*, March 18, 2019, https://www.wired.com/story/people-trying-make -internet-recommendations-less-toxic/. On the broader political, economic, and social implications of this new form of power, see: Shoshana Zuboff, *Surveillance Capitalism: The Fight for a Human Future at the New Frontier of Power* (New York: Public Affairs, 2019).

23. Steven Levy, "Inside Facebook's AI Machine," *Wired*, February 23, 2017, https:// www.wired.com/2017/02/inside-facebooks-ai-machine/.

24. Jeff Horwitz and Deepa Seetharaman, "Facebook Executives Shut Down Efforts to Make the Site Less Divisive," *Wall Street Journal*, May 26, 2020, https://www.wsj .com/articles/facebook-knows-it-encourages-division-top-executives-nixed-solutions -11590507499.

25. Buchanan, *The Hacker and the State*, 229–236.

26. Horwitz and Seetharaman, "Facebook Executives Shut Down Efforts to Make the Site Less Divisive."

27. Casey Newton, "Mark in the Middle," *Verge*, September 23, 2020, https://www .theverge.com/21444203/facebook-leaked-audio-zuckerberg-trump-pandemic-blm.

28. Ryan Mac and Craig Silverman, "Facebook Quietly Suspended Political Group Recommendations Ahead of the US Presidential Election," *BuzzFeed*, October 30, 2020, https://www.buzzfeednews.com/article/ryanmac/facebook-suspended-group -recommendations-election; Leon Yin and Alfred Ng, "Facebook Said It Would Stop Pushing Users to Join Partisan Political Groups. It Didn't," *Markup*, January 19, 2021, https://themarkup.org/citizen-browser/2021/01/19/facebook-said-it-would-stop -pushing-users-to-join-partisan-political-groups-it-didnt.

29. Kaili Lambe and Becca Ricks, "The Basics on Microtargeting and Political Ads on Facebook," Mozilla Foundation, January 14, 2020, https://foundation.mozilla.org /en/blog/basics-microtargeting-and-political-ads-facebook/.

30. It is illegal for noncitizens to purchase ads designed to influence American elections. Grand Jury for the District of Columbia, *Indictment: United States of America v. Internet Research Agency LLC* (United States District Court for the District of Columbia,

February 16, 2018), 19–20, https://www.justice.gov/file/1035477/download. For copies of the ads, see Permanent Select Committee on Intelligence, "Social Media Advertisements," United States House of Representatives, May 10, 2018, https://intelligence.house.gov/social-media-content/social-media-advertisements.htm.

31. Colin Stretch, *Testimony of Colin Stretch, General Counsel, Facebook* (Hearing before the United States Senate Committee on the Judiciary Subcommittee on Crime and Terrorism, October 31, 2017), https://www.judiciary.senate.gov/imo/media/doc/10-31-17%20Stretch%20Testimony.pdf; Mike Isaac and Daisuke Wakabayashi, "Russian Influence Reached 126 Million Through Facebook Alone," *New York Times*, October 30, 2017, https://www.nytimes.com/2017/10/30/technology/facebook-google-russia.html; Andrew M. Guess, Brendan Nyhan, and Jason Reifler, "Exposure to Untrustworthy Websites in the 2016 US Election," *Nature Human Behaviour* 4, no. 5 (2020): 472–480.

32. "April 2020 Coordinated Inauthentic Behavior Report," Facebook News, May 5, 2020, https://about.fb.com/news/2020/05/april-cib-report/; Emily Dreyfuss and Issie Lapowsky, "Facebook Is Changing News Feed (Again) to Stop Fake News," *Wired*, April 10, 2019, https://www.wired.com/story/facebook-click-gap-news-feed-changes/; Kevin Roose, "Facebook Reverses Postelection Algorithm Changes That Boosted News from Authoritative Sources," *New York Times*, December 16, 2020, https://www.nytimes.com/2020/12/16/technology/facebook-reverses-postelection-algorithm-changes-that-boosted-news-from-authoritative-sources.html.

33. *Facebook's Civil Rights Audit—Final Report* (Facebook, July 8, 2020), 56, https://about.fb.com/wp-content/uploads/2020/07/Civil-Rights-Audit-Final-Report.pdf.

34. Craig Silverman, Ryan Mac, and Jane Lytvynenko, "Facebook Knows It Was Used To Help Incite the Capitol Insurrection," *BuzzFeed*, April 22, 2021, https://www.buzzfeednews.com/article/craigsilverman/facebook-failed-stop-the-steal-insurrection; Joan Donovan, "From 'Get Big Fast' to 'Move Fast and Break Things' and Back Again," Statement of Joan Donovan at Hearing on "Algorithms and Amplification: How Social Media Platforms' Design Choices Shape Our Discourse and Our Minds," Senate Committee on the Judiciary, Subcommittee on Privacy, Technology, and the Law, April 27, 2021, https://www.judiciary.senate.gov/imo/media/doc/Donovan%20Testimony%20(updated).pdf.

35. James Vincent, "Facebook Is Now Using AI to Sort Content for Quicker Moderation," *Verge*, November 13, 2020, https://www.theverge.com/2020/11/13/21562596/facebook-ai-moderation.

36. "YouTube for Press," YouTube, August 22, 2020, https://web.archive.org/web/20200822001707/https://www.youtube.com/about/press/.

37. Paul Covington, Jay Adams, and Emre Sargin, "Deep Neural Networks for YouTube Recommendations," in *Proceedings of the 10th ACM Conference on Recommender Systems* (New York: Association for Computing Machinery, 2016), 191–198.

38. Paresh Dave and Christopher Bing, "Russian Disinformation on YouTube Draws Ads, Lacks Warning Labels: Researchers," *Reuters*, June 7, 2019, https://www.reuters.com/article/us-alphabet-google-youtube-russia-idUSKCN1T80JP.

39. Drew Harwell and Craig Timberg, "Youtube Recommended a Russian Media Site Thousands of Times for Analysis of Mueller's Report, a Watchdog Group Says," *Washington Post*, April 26, 2019, https://www.washingtonpost.com/technology /2019/04/26/youtube-recommended-russian-media-site-above-all-others-analysis -mueller-report-watchdog-group-says/.

40. Harwell and Timberg, "YouTube Recommended a Russian Media Site Thousands of Times for Analysis of Mueller's Report."

41. Harwell and Timberg, "YouTube Recommended a Russian Media Site Thousands of Times for Analysis of Mueller's Report."

42. Paul Lewis, "'Fiction Is Outperforming Reality': How YouTube's Algorithm Distorts Truth," *Guardian*, February 2, 2018, http://www.theguardian.com/technology /2018/feb/02/how-youtubes-algorithm-distorts-truth.

43. Lewis, "'Fiction Is Outperforming Reality.'"

44. Daniel Howley, "Google's Public Response to Russian Trolls Is Very Different from Facebook's," *Yahoo! Finance*, December 13, 2018, https://finance.yahoo.com /news/google-facebook-different-public-responses-russian-trolls-185140923.html.

45. "John Grierson, Screen Pioneer Who Made Documentaries Dies," *New York Times*, February 21, 1972, https://www.nytimes.com/1972/02/21/archives/john-grierson -screen-pneer-who-madedocumentaries-dles-sped-tois-or.html; "The Art of Propaganda, Part One," *Rewind*, CBC Radio, November 3, 2016, https://www.cbc.ca/radio /rewind/the-art-of-propaganda-part-one-1.3829270.

46. "Malicious Actors Almost Certainly Will Leverage Synthetic Content for Cyber and Foreign Influence Operations," Federal Bureau of Investigation, March 10, 2021, https://www.documentcloud.org/documents/20509703-fbipin-3102021.

47. Daniel R. Coats, *Worldwide Threat Assessment of the US Intelligence Community* (US Senate Select Committee on Intelligence, January 29, 2019), 7, https://www.dni .gov/files/ODNI/documents/2019-ATA-SFR---SSCI.pdf.

48. "Fake Cluster Boosts Huawei"; Adam Satariano, "Inside a Pro-Huawei Influence Campaign," *New York Times*, January 29, 2021, https://www.nytimes.com/2021 /01/29/technology/commercial-disinformation-huawei-belgium.html.

49. Robert Chesney, Danielle Citron, and Hany Farid, "All's Clear for Deepfakes: Think Again," *Lawfare*, May 11, 2020, https://www.lawfareblog.com/alls-clear -deepfakes-think-again; Robert Chesney and Danielle Citron, "Deep Fakes: A Looming Challenge for Privacy, Democracy, and National Security," *California Law Review* 107 (2019): 1753.

50. Hannah Arendt, *The Origins of Totalitarianism* (Boston: Houghton Mifflin Harcourt, 1976), 350.

51. George Orwell, *1984* (Boston: Houghton Mifflin Harcourt, 1983), ch. 7.

52. Sarah Mervosh, "Distorted Videos of Nancy Pelosi Spread on Facebook and Twitter, Helped by Trump," *New York Times*, May 24, 2019, https://www.nytimes.com /2019/05/24/us/politics/pelosi-doctored-video.html; Emily Stewart, "A Fake Viral

Video Makes Nancy Pelosi Look Drunk. Facebook Won't Take It down," *Vox*, May 24, 2019, https://www.vox.com/recode/2019/5/24/18638822/nancy-pelosi-doctored-video -drunk-facebook-trump; Hannah Denham, "Another Fake Video of Pelosi Goes Viral on Facebook," *Washington Post*, August 3, 2020, https://www.washingtonpost.com /technology/2020/08/03/nancy-pelosi-fake-video-facebook/.

53. Craig Silverman, "This Analysis Shows How Viral Fake Election News Stories Outperformed Real News on Facebook," *BuzzFeed*, November 16, 2016, https://www .buzzfeednews.com/article/craigsilverman/viral-fake-election-news-outperformed -real-news-on-facebook.

54. Rand Waltzman, *Testimony: The Weaponization of Information* (United States Senate Committee on Armed Services, April 27, 2017), https://www.armed-services .senate.gov/imo/media/doc/Waltzman_04-27-17.pdf. See also Cori E. Dauber, "The Truth Is Out There: Responding to Insurgent Disinformation and Deception Opera- tions," *Military Review* 89, no. 1 (January–February 2009).

55. Waltzman, "Testimony: The Weaponization of Information."

56. Interview with the authors, February 27, 2020.

57. Interview with the authors.

58. "Adobe Research and UC Berkeley: Detecting Facial Manipulations in Adobe Photoshop," Adobe blog, June 14, 2019, https://blog.adobe.com/en/2019/06/14 /adobe-research-and-uc-berkeley-detecting-facial-manipulations-in-adobe-photoshop .html.

59. John Keller, "Disinformation Falsified Media Algorithms," Military and Aero- space Electronics, August 24, 2020, https://www.militaryaerospace.com/trusted -computing/article/14182166/disinformation-falsified-media-algorithms.

60. Interview with the authors.

61. "Creating a Data Set and a Challenge for Deepfakes," Facebook blog, September 5, 2019, https://ai.facebook.com/blog/deepfake-detection-challenge/. For a sample of other technical work, see Shruti Agarwal et al., "Detecting Deep-Fake Videos from Phoneme-Viseme Mismatches," in *Proceedings of the IEEE/CVF Conference on Com- puter Vision and Pattern Recognition Workshops* (2020): 660–661.

62. Will Douglas Heaven, "Facebook Just Released a Database of 100,000 Deepfakes to Teach AI How to Spot Them," *MIT Technology Review*, June 12, 2020, https://www .technologyreview.com/2020/06/12/1003475/facebooks-deepfake-detection -challenge-neural-network-ai/.

63. Interview with the authors.

64. Interview with the authors.

65. Tripp Mickle, "Secretive Apple Tries to Open Up on Artificial Intelligence," *Wall Street Journal*, September 3, 2017, https://www.wsj.com/articles/secretive -apple-tries-to-open-up-on-artificial-intelligence-1504436401.

66. Daniel Ziegler et al., "Fine-Tuning GPT-2 from Human Preferences," OpenAI, September 19, 2019, https://openai.com/blog/fine-tuning-gpt-2/.

67. Radford et al., "Better Language Models and Their Implications."

68. "Elon Musk's OpenAI Builds Artificial Intelligence So Powerful It Must Be Kept Locked Up for the Good of Humanity," *Business Fast*, February 15, 2019, https://www.businessfast.co.uk/elon-musks-openai-builds-artificial-intelligence-so-powerful-it-must-be-kept-locked-up-for-the-good-of-humanity/.

69. Richard Socher, "But their perplexity on wikitext-103 is 0.8 lower than previous sota. So it's dangerous now. ☺ In other news. Copy and pasting and Photoshop are existential threats to humanity. PS: Love language models. Love multitask learning. Great work. Dislike the hype and fear mongering," Twitter, February 14, 2019, 6:53 p.m., https://twitter.com/RichardSocher/status/1096195833865789442.

70. Yann LeCun, "<trolling-joking> Every new human can potentially be used to generate fake news, disseminate conspiracy theories, and influence people. Should we stop making babies then? </trolling-joking>," Twitter, February 19, 2019, 10:18 a.m., https://twitter.com/ylecun/status/1097878015667851266.

71. Some of the early OpenAI releases misstated the number of parameters.

72. House Permanent Select Committee on Intelligence, "Open Hearing on Deepfakes and Artificial Intelligence," YouTube video, June 13, 2019, 2:18:20, https://www.youtube.com/watch?v=tdLS9MlIWOk.

73. Alexander Graves and Kelly Clancy, "Unsupervised Learning: The Curious Pupil," DeepMind, June 25, 2019, https://deepmind.com/blog/article/unsupervised-learning.

74. Claire Leibowicz, Steven Adler, and Peter Eckersley, "When Is It Appropriate to Publish High-Stakes AI Research?," Partnership on AI, April 2, 2019, https://www.partnershiponai.org/when-is-it-appropriate-to-publish-high-stakes-ai-research/.

75. Aviv Ovadya and Jess Whittlestone, "Reducing Malicious Use of Synthetic Media Research: Considerations and Potential Release Practices for Machine Learning," arXiv, July 25, 2019, http://arxiv.org/abs/1907.11274; Rowan Zellers et al., "Defending Against Neural Fake News," in *Advances in Neural Information Processing Systems* 32 (2019), 9054–9065; Miles Brundage et al., "Toward Trustworthy AI Development: Mechanisms for Supporting Verifiable Claims," arXiv, April 15, 2020, http://arxiv.org/abs/2004.07213.

76. Irene Solaiman, Jack Clark, and Miles Brundage, "GPT-2: 1.5B Release," OpenAI, November 5, 2019, https://openai.com/blog/gpt-2-1-5b-release/.

77. Brown et al., "Language Models Are Few-Shot Learners."

78. An API, or an application programming interface, is a mechanism to issue commands or receive information from a piece of software.

79. Greg Brockman, Mira Murati, and Peter Welinder, "OpenAI API," OpenAI, June 11, 2020, https://openai.com/blog/openai-api/.

80. Ben Buchanan et al., *Truth, Lies, and Automation: How Language Models Could Change Disinformation* (Georgetown University Center for Security and Emerging Technology, May 2021).

81. Toby Shevlane and Allan Dafoe, "The Offense-Defense Balance of Scientific Knowledge: Does Publishing AI Research Reduce Misuse?," in *Proceedings of the AAAI/ACM Conference on AI, Ethics, and Society* (New York: Association for Computing Machinery, 2020), 173–179.

82. For a paper from a group replicating GPT-3, see Leo Gao et al., "The Pile: An 800GB Dataset of Diverse Text for Language Modeling," arXiv, December 31, 2020, http://arxiv.org/abs/2101.00027.

83. "Getting Started with NeurIPS 2020," Neural Information Processing Systems Conference, February 19, 2020, https://neuripsconf.medium.com/getting-started-with-neurips-2020-e350f9b39c28.

84. Inioluwa Deborah Raji et al., "Closing the AI Accountability Gap: Defining an End-to-End Framework for Internal Algorithmic Auditing," arXiv, January 3, 2020, http://arxiv.org/abs/2001.00973.

85. Halvar Flake, "Two Small Notes on the 'Malicious Use of AI' Report," ADD / XOR / ROL, February 21, 2018, https://addxorrol.blogspot.com/2018/02/two-small-notes-on-malicious-use-of-ai.html.

86. G. H. Hardy, *A Mathematician's Apology* (Alberta: University of Alberta Mathematical Sciences Society, 2005), 49.

87. Hardy, *A Mathematician's Apology*, 44.

88. Michelle Wagner, "The Inside Scoop on Mathematics at the NSA," *Math Horizons* 13, no. 4 (2006): 20–23.

CHAPTER 9

1. *United States-Soviet Relations Hearings before the Committee on Foreign Relations, United States Senate, Ninety-Eighth Congress, Part 1* (Washington, DC: U.S. Government Printing Office, 1983), 49.

2. Simon Miles, "The War Scare That Wasn't: Able Archer 83 and the Myths of the Second Cold War," *Journal of Cold War Studies* 22, no. 3 (2020): 91–93; Christopher M. Andrew and Oleg Gordievsky, *Comrade Kryuchkov's Instructions: Top Secret Files on KGB Foreign Operations, 1975–1985* (Stanford, CA: Stanford University Press, 1993), 67.

3. Benjamin B. Fischer, "KGB Cable No. 373/PR/5217.02.83, 'Comrade Yermakov to A. V. Guk: Permanent Operational Assignment to Uncover NATO Preparations for a Nuclear Missile Attack on the USSR,'" Central Intelligence Agency, July 7, 2008, https://web.archive.org/web/20200617053757/https://www.cia.gov/library/center-for-the-study-of-intelligence/csi-publications/books-and-monographs/a-cold-war-conundrum/source.htm.

4. Benjamin B. Fischer, "Appendix A: RYAN and the Decline of the KGB," Center for the Study of Intelligence, Central Intelligence Agency, 1997, https://web.archive.org/web/20060802003243/https://www.cia.gov/csi/monograph/coldwar/source.htm.

5. Ronald Reagan, "'Evil Empire' Speech Text," Voices of Democracy, March 8, 1983, https://voicesofdemocracy.umd.edu/reagan-evil-empire-speech-text/.

6. Nate Jones and David E. Hoffman, "Newly Released Documents Shed Light on 1983 War Scare with Soviets," Washington Post, February 17, 2021, https://www.washingtonpost.com/national-security/soviet-nuclear-war-able-archer/2021/02/17/711fa9e2-7166-11eb-93be-c10813e358a2_story.html.

7. Geoffrey Wiseman, Concepts of Non-Provocative Defence: Ideas and Practices in International Security (London: Palgrave, 2002), 250.

8. Miles, "The War Scare That Wasn't"; David E. Hoffman, "In 1983 'War Scare,' Soviet Leadership Feared Nuclear Surprise Attack by U.S.," Washington Post, October 24, 2015, https://www.washingtonpost.com/world/national-security/in-1983-war-scare-soviet-leadership-feared-nuclear-surprise-attack-by-us/2015/10/24/15a289b4-7904-11e5-a958-d889faf561dc_story.html.

9. Robert Jervis, "Cooperation under the Security Dilemma," World Politics 30, no. 2 (1978): 167–214.

10. "Panama Invasion: The US Operation That Ousted Noriega," BBC, December 20, 2019, https://www.bbc.com/news/world-latin-america-50837024; Ezer Vierba, "Panama's Stolen Archive," NACLA, September 28, 2014, https://nacla.org/article/panama%E2%80%99s-stolen-archive.

11. For a period of time in the 1960s, the Soviets developed the Fractional Orbital Bombardment System, designed to launch nuclear weapons from space. Such a capability fell out of favor for a variety of reasons, including that it was less useful than submarine-launched missiles. See Asif A. Siddiqi, "The Soviet Fractional Orbiting Bombardment System (FOBS): A Short Technical History," Quest: The History of Spaceflight Quarterly 7, no. 4 (2000): 22–32.

12. Pavel Aksenov, "Stanislav Petrov: The Man Who May Have Saved the World," BBC, September 26, 2013, https://www.bbc.com/news/world-europe-24280831.

13. David Hoffman, "'I Had A Funny Feeling in My Gut,'" Washington Post, February 10, 1999, https://www.washingtonpost.com/wp-srv/inatl/longterm/coldwar/shatter021099b.htm.

14. Aksenov, "Stanislav Petrov."

15. James M. Acton, "Escalation through Entanglement: How the Vulnerability of Command-and-Control Systems Raises the Risks of an Inadvertent Nuclear War," International Security 43, no. 1 (Summer 2018): 56–99, https://direct.mit.edu/isec/article/43/1/56/12199/Escalation-through-Entanglement-How-the.

16. Nicholas Thompson, "Inside the Apocalyptic Soviet Doomsday Machine," Wired, September 21, 2009, https://www.wired.com/2009/09/mf-deadhand/.

17. Robert R. Everett, "Semi-Automatic Ground Environment," Federation of American Scientists, June 29, 1999, https://fas.org/nuke/guide/usa/airdef/sage.htm.

18. "Manhattan Project: CTBTO Preparatory Commission," Comprehensive Nuclear-Test-Ban Treaty Organization, accessed July 1, 2020, https://www.ctbto.org/nuclear-testing/history-of-nuclear-testing/manhattan-project/; Sebastian Anthony, "Inside

IBM's $67 Billion SAGE, the Largest Computer Ever Built," *Extreme Tech*, March 28, 2013, https://www.extremetech.com/computing/151980-inside-ibms-67-billion-sage -the-largest-computer-ever-built.

19. Herman Kahn, *On Escalation: Metaphors and Scenarios* (Piscataway, NJ: Transaction Publishers, 2009), 11. For more on how this idea applies to AI, see Ryan Fedasiuk, "Second, there is the Gen. Jack Ripper model of deliberate abuse. What is to stop someone from artificially triggering the glass- or flash-sensors that would be used in a Dead Hand system?" Twitter, August 16, 2019, 10:16 a.m., https://twitter .com/RyanFedasiuk/status/1162397697216671745.

20. Lora Saalman, "Integration of Neural Networks into Hypersonic Glide Vehicles," *Impact of Artificial Intelligence on Strategic Stability and Nuclear Risk: East Asian Perspectives* 2 (October 2019): 24–25.

21. Keir A. Lieber and Daryl G. Press, "The New Era of Counterforce: Technological Change and the Future of Nuclear Deterrence," *International Security* 41, no. 4 (2017): 9–49; Michael C. Horowitz, Paul Scharre, and Alexander Velez-Green, "A Stable Nuclear Future? The Impact of Autonomous Systems and Artificial Intelligence," arXiv, December 11, 2019, http://arxiv.org/abs/1912.05291.

22. Phil Stewart, "Deep in the Pentagon, a Secret AI Program to Find Hidden Nuclear Missiles," *Reuters*, June 5, 2018, https://www.reuters.com/article/us-usa-pentagon -missiles-ai-insight-idUSKCN1J114J; Li Xiang, "Artificial Intelligence and Its Impact on Weaponization and Arms Control," *Impact of Artificial Intelligence on Strategic Stability and Nuclear Risk: East Asian Perspectives* 2 (October 2019): 13–14.

23. Michael C. Horowitz, "When Speed Kills: Autonomous Weapon Systems, Deterrence, and Stability," *Journal of Strategic Studies* 42, no. 6 (2019): 21, https://www .tandfonline.com/doi/abs/10.1080/01402390.2019.1621174?journalCode=fjss20; Sydney J. Freedberg, "Transparent Sea: The Unstealthy Future of Submarines," *Breaking Defense*, January 22, 2015, https://breakingdefense.com/2015/01/transparent-sea -the-unstealthy-future-of-submarines/; Tuneer Mukherjee, "Securing the Maritime Commons: The Role of Artificial Intelligence in Naval Operations," *OFR Occasional Paper* 159 (July 2018); Andrew Reddie and Bethany Goldblum, "Unmanned Underwater Vehicle (UUV) Systems for Submarine Detection," On the Radar, July 29, 2019, https://ontheradar.csis.org/issue-briefs/unmanned-underwater-vehicle-uuv-systems -for-submarine-detection-a-technology-primer/.

24. Erik Gartzke and Jon R. Lindsay, "Thermonuclear Cyberwar," *Journal of Cybersecurity* 3, no. 1 (2017): 37–48.

25. Adam Lowther and Curtis McGiffin, "America Needs a 'Dead Hand,'" *War on the Rocks*, August 16, 2019, https://warontherocks.com/2019/08/america-needs-a-dead -hand/; Luke O'Brien, "Whither Skynet? An American 'Dead Hand' Should Remain a Dead Issue," *War on the Rocks*, September 11, 2019, https://warontherocks.com /2019/09/whither-skynet-an-american-dead-hand-should-remain-a-dead-issue/.

26. Lora Saalman, "Fear of False Negatives: AI and China's Nuclear Posture," *Bulletin of the Atomic Scientists*, April 24, 2018, https://thebulletin.org/2018/04/fear-of -false-negatives-ai-and-chinas-nuclear-posture/.

27. James M. Acton, "Escalation through Entanglement," *International Security* 43, no. 1 (Summer 2018): 56–99; Lora Saalman, "Introduction," *Impact of Artificial Intelligence on Strategic Stability and Nuclear Risk: East Asian Perspectives* 2 (October 2019): 5.

28. Rebecca Hersman, "Wormhole Escalation in the New Nuclear Age," *Texas National Security Review*, Summer 2020, https://tnsr.org/2020/07/wormhole-escalation-in-the -new-nuclear-age/; Alexey Arbatov, Vladimir Dvorkin, and Petr Topychkanov, "Entanglement as a New Security Threat: A Russian Perspective," Carnegie Endowment for International Peace, November 8, 2017, https://carnegieendowment.org/2017/11/08 /entanglement-as-new-security-threat-russian-perspective-pub-73163.

29. Yuna H. Wong et al., *Deterrence in the Age of Thinking Machines* (Santa Monica, CA: RAND Corporation, 2020), https://apps.dtic.mil/sti/citations/AD1090316.

30. Roger Fisher, "Preventing Nuclear War," *Bulletin of the Atomic Scientists* 37, no. 3 (1981): 11–17.

31. Interview with the authors, June 26, 2020.

32. "The Intermediate-Range Nuclear Forces (INF) Treaty at a Glance," Arms Control Association, August 2019, https://www.armscontrol.org/factsheets/INFtreaty; "Press Briefing with Andrea L. Thompson Under Secretary for Arms Control And International Security Affairs," Department of State, January 16, 2019, https://www .state.gov/press-briefing-with-andrea-l-thompson-under-secretary-for-arms-control-and -international-security-affairs/.

33. Interview with the authors.

34. Arvid Schors, "Trust and Mistrust and the American Struggle for Verification of the Strategic Arms Limitation Talks, 1969–1979," in *Trust, but Verify: The Politics of Uncertainty and the Transformation of the Cold War Order, 1969–1991*, ed. Martin Klimke, Reinhild Kreis, and Christian F. Ostermann (Palo Alto, CA: Stanford University Press, 2016), 85–101; Rose Gottemoeller, "U.S.–Russian Nuclear Arms Control Negotiations—A Short History," American Foreign Service Association, May 2020, https://www.afsa.org/us-russian-nuclear-arms-control-negotiations-short-history.

35. Michael Horowitz and Paul Scharre, "AI and International Stability: Risks and Confidence-Building Measures," Center for a New American Security, January 12, 2021, https://www.cnas.org/publications/reports/ai-and-international-stability-risks -and-confidence-building-measures; Andrew Imbrie and Elsa B. Kania, *AI Safety, Security, and Stability Among Great Powers* (Georgetown University Center for Security and Emerging Technology, December 2019), https://cset.georgetown.edu/wp-content /uploads/AI-Safety-Security-and-Stability-Among-the-Great-Powers.pdf.

36. Interview with the authors.

37. Interview with the authors.

38. Interview with the authors.

39. Andrew Imbrie, "Competitive Strategies for Democracy in the Age of AI," German Marshall Fund, June 30, 2020, https://securingdemocracy.gmfus.org /competitive-strategies-for-democracy-in-the-age-of-ai/; "U.S. Collective Defense

Arrangements," Department of State, January 20, 2017, https://2009-2017.state.gov
/s/l/treaty/collectivedefense/index.htm; Mira Rapp-Hooper, *Shields of the Republic:
The Triumph and Peril of America's Alliances* (Cambridge, MA: Harvard University
Press, 2020); Michael Beckley, "The Myth of Entangling Alliances: Reassessing the
Security Risks of U.S. Defense Pacts," *International Security* 39, no. 4 (2015): 7–48.

40. Rapp-Hooper, *Shields of the Republic,* 47–77; Hal Brands and Peter D. Feaver,
"What Are America's Alliances Good For?," *Parameters: Journal of the US Army War
College* 47, no. 2 (2017), http://search.proquest.com/openview/f9a0e0021171ed143
cf41f3c3b3d347e/1?pq-origsite=gscholar&cbl=32439; Andrew Imbrie, *Power on the
Precipice: The Six Choices America Faces in a Turbulent World* (New Haven, CT: Yale
University Press, 2020), 87–93.

41. Interview with the authors.

42. Interview with the authors.

43. Interview with the authors, June 24, 2020.

44. Interview with the authors.

45. Interview with the authors.

46. Interview with the authors.

47. *The AIM Initiative* (Office of the Director of National Intelligence, January 16,
2019), https://www.dni.gov/files/ODNI/documents/AIM-Strategy.pdf; Amy Zegart
and Michael Morell, "Spies, Lies, and Algorithms," *Foreign Affairs*, May–June 2019,
https://fsi-live.s3.us-west-1.amazonaws.com/s3fs-public/zegartmorell.pdf.

48. Interview with the authors.

49. Colin Clark, "Cardillo: 1 Million Times More GEOINT Data in 5 Years," *Breaking
Defense*, June 5, 2017, https://breakingdefense.com/2017/06/cardillo-1-million-times
-more-geoint-data-in-5-years/; Theresa Hitchens, "IC Must Embrace Public Data to
Use AI Effectively: Sue Gordon," *Breaking Defense*, September 25, 2019, https://
breakingdefense.com/2019/09/ic-must-embrace-public-data-to-use-ai-effectively-sue
-gordon/.

50. Jenna McLaughlin, "Artificial Intelligence Will Put Spies out of Work, Too," *Foreign Policy*, June 9, 2017, https://foreignpolicy.com/2017/06/09/artificial-intelligence
-will-put-spies-out-of-work-too/.

51. Interview with the authors.

52. Interview with the authors. Gordon was speaking generally and not referring to
any specific program or agency.

53. Ava Kofman, "Forget about Siri and Alexa—When It Comes to Voice Identification, the 'NSA Reigns Supreme,'" *Intercept*, January 19, 2018, https://theintercept.com
/2018/01/19/voice-recognition-technology-nsa/.

54. Charles A. Shoniregun and Stephen Crosier, eds., "Research Overview and Biometric Technologies," in *Securing Biometrics Applications* (Boston: Springer, 2008), 16.

55. Kofman, "Forget about Siri and Alexa."

56. Kofman, "Forget about Siri and Alexa."

57. Dahlia Peterson, *Designing Alternatives to China's Repressive Surveillance State* (Georgetown University Center for Security and Emerging Technology, October 2020), https://cset.georgetown.edu/wp-content/uploads/CSET-Designing-Alternatives -to-Chinas-Surveillance-State.pdf; Patrick Tucker, "Spies Like AI: The Future of Artifi- cial Intelligence for the US Intelligence Community," *Defense One*, January 27, 2020, https://www.defenseone.com/technology/2020/01/spies-ai-future-artificial-intelligence -us-intelligence-community/162673/.

58. In part due to concerns about iFlyTek's work with the Chinese government, MIT cut ties with the company in 2020. "China: Voice Biometric Collection Threatens Privacy," Human Rights Watch, October 22, 2017, https://www.hrw.org/news/2017 /10/22/china-voice-biometric-collection-threatens-privacy; Wu Ji et al., "System and method for realizing audio file repeating pattern finding," CN:103440270:A, patent filed August 2, 2013, and issued December 11, 2013, https://patentimages.storage .googleapis.com/09/80/4c/0b1806a5be5c18/CN103440270A.pdf; Will Knight, "MIT Cuts Ties with a Chinese AI Firm Amid Human Rights Concerns," *Wired*, April 21, 2020, https://www.wired.com/story/mit-cuts-ties-chinese-ai-firm-human-rights/.

59. Tucker, "Spies Like AI."

60. Interview with the authors.

61. Interpreting intentions has also been the subject of substantial academic study. For example, see Keren Yarhi-Milo, *Knowing the Adversary: Leaders, Intelligence, and Assessment of Intentions in International Relations* (Princeton, NJ: Princeton University Press, 2014).

62. Interview with the authors.

CHAPTER 10

1. Austin Ramzy and Chris Buckley, "'Absolutely No Mercy': Leaked Files Expose How China Organized Mass Detentions of Muslims," *New York Times*, November 16, 2019, https://www.nytimes.com/interactive/2019/11/16/world/asia/china-xinjiang -documents.html. This story was part of a broader global partnership. Bethany Allen-Ebrahimian, "Exposed: China's Operating Manuals for Mass Internment and Arrest by Algorithm," International Consortium of Investigative Journalists, November 24, 2019, https://www.icij.org/investigations/china-cables/exposed-chinas -operating-manuals-for-mass-internment-and-arrest-by-algorithm/.

2. Ramzy and Buckley, "'Absolutely No Mercy.'"

3. Gerry Shih, "Digital Police State Shackles Chinese Minority," *AP News*, December 17, 2017, https://apnews.com/1ec5143fe4764a1d8ea73ce4a3e2c570/AP-Exclusive: -Digital-police-state-shackles-Chinese-minority.

4. Lily Kuo, "China Footage Reveals Hundreds of Blindfolded and Shackled Prison- ers," *Guardian*, September 23, 2019, https://www.theguardian.com/world/2019/sep /23/china-footage-reveals-hundreds-of-blindfolded-and-shackled-prisoners-uighur; Paul Mozur, "One Month, 500,000 Face Scans: How China Is Using A.I. to Profile a

Minority," *New York Times*, April 14, 2019, https://www.nytimes.com/2019/04/14 /technology/china-surveillance-artificial-intelligence-racial-profiling.html.

5. Sheena Chestnut Greitens, Myunghee Lee, and Emir Yazici, "Counterterrorism and Preventive Repression: China's Changing Strategy in Xinjiang," *International Security* 44, no. 3 (2020): 9–47, https://www.belfercenter.org/publication/counter terrorism-and-preventive-repression-chinas-changing-strategy-xinjiang; James Millward and Dahlia Peterson, *China's System of Oppression in Xinjiang: How It Developed and How to Curb It* (Brookings Institution, September 2020), https://www.brookings .edu/wp-content/uploads/2020/09/FP_20200914_china_oppression_xinjiang _millward_peterson.pdf.

6. "China: Big Data Fuels Crackdown in Minority Region," Human Rights Watch, February 26, 2018, https://www.hrw.org/news/2018/02/26/china-big-data-fuels -crackdown-minority-region; Ross Andersen, "The Panopticon Is Already Here," *Atlantic*, July 29, 2020, https://www.theatlantic.com/magazine/archive/2020/09 /china-ai-surveillance/614197/; Yael Grauer, "Revealed: Massive Chinese Police Database," *Intercept*, January 29, 2021, https://theintercept.com/2021/01/29/china -uyghur-muslim-surveillance-police/.

7. Paul Mozur, "China's Surveillance State Sucks Up Data. U.S. Tech Is Key to Sorting It," *New York Times*, November 22, 2020, https://www.nytimes.com/2020/11/22 /technology/china-intel-nvidia-xinjiang.html. For example, see Eva Dou and Drew Harwell, "Huawei Worked on Several Surveillance Systems Promoted to Identify Ethnicity, Documents Show," *Washington Post*, December 12, 2020, https://www .washingtonpost.com/technology/2020/12/12/huawei-uighurs-identify/; Drew Harwell and Eva Dou, "Huawei Tested AI Software That Could Recognize Uighur Minorities and Alert Police, Report Says," *Washington Post*, December 8, 2020, https:// www.washingtonpost.com/technology/2020/12/08/huawei-tested-ai-software-that -could-recognize-uighur-minorities-alert-police-report-says/; Raymond Zhong, "As China Tracked Muslims, Alibaba Showed Customers How They Could, Too," *New York Times*, December 16, 2020, https://www.nytimes.com/2020/12/16/technology /alibaba-china-facial-recognition-uighurs.html.

8. Sheena Chestnut Greitens, "Dealing with Demand for China's Global Surveillance Exports," (Brookings Institution, April 2020), https://www.brookings.edu /wp-content/uploads/2020/04/FP_20200428_china_surveillance_greitens_v3.pdf.

9. Stephen Kafeero, "Uganda Is Using Huawei's Facial Recognition Tech to Crack Down on Dissent after Anti-Government Protests," *Quartz*, November 27, 2020, https://qz.com/africa/1938976/uganda-uses-chinas-huawei-facial-recognition-to -snare-protesters/.

10. Amy Hawkins, "Beijing's Big Brother Tech Needs African Faces," *Foreign Policy*, July 24, 2018, https://foreignpolicy.com/2018/07/24/beijings-big-brother-tech-needs -african-faces/; Paul Mozur, Jonah M. Kessel, and Melissa Chan, "Made in China, Exported to the World: The Surveillance State," *New York Times*, April 24, 2019, https://www.nytimes.com/2019/04/24/technology/ecuador-surveillance-cameras -police-government.html.

11. Harari, "Why Technology Favors Tyranny."

12. Jaron Lanier and Glen Weyl, "AI Is an Ideology, Not a Technology," *Wired*, March 15, 2020, https://www.wired.com/story/opinion-ai-is-an-ideology-not-a -technology/.

13. "The Global AI Talent Tracker," MacroPolo, accessed July 1, 2020, https:// macropolo.org/digital-projects/the-global-ai-talent-tracker/.

14. Remco Zwetsloot et al., *Keeping Top AI Talent in the United States* (Georgetown University Center for Security and Emerging Technology, December 2019), https:// cset.georgetown.edu/wp-content/uploads/Keeping-Top-AI-Talent-in-the-United-States .pdf; Will Hunt and Remco Zwetsloot, *The Chipmakers: U.S. Strengths and Priorities for the High-End Semiconductor Workforce* (Georgetown University Center for Security and Emerging Technology, September 2020), https://cset.georgetown.edu/wp -content/uploads/CSET-The-Chipmakers.pdf; Remco Zwetsloot, Roxanne Heston, and Zachary Arnold, *Strengthening the US AI Workforce* (Georgetown University Center for Security and Emerging Technology, September 2019), https://cset .georgetown.edu/wp-content/uploads/CSET_US_AI_Workforce.pdf.

15. Zwetsloot et al., *Keeping Top AI Talent in the United States*.

16. For an example of a more recent innovation that occurred in China and for discussion of determining who benefits, see Sheehan, "Who Benefits from American AI Research in China?"

17. Zachary Arnold, *Canada's Skilled Immigration System Increasingly Draws Talent from the United States* (Georgetown University Center for Security and Emerging Technology, July 14, 2020), https://cset.georgetown.edu/wp-content/uploads/CSET -Canadas-Skilled-Immigration-System-Increasingly-Draws-Talent-from-the-United -States.pdf.

18. Ryan Fedasiuk and Jacob Feldgoise, *The Youth Thousand Talents Plan and China's Military* (Georgetown University Center for Security and Emerging Technology, August 2020), https://cset.georgetown.edu/wp-content/uploads/CSET-Youth-Thousand -Talents-Plan-and-Chinas-Military.pdf; Cheng Ting-Fang, "China Hires over 100 TSMC Engineers in Push for Chip Leadership," *Nikkei*, August 12, 2020, https://asia .nikkei.com/Business/China-tech/China-hires-over-100-TSMC-engineers-in-push-for -chip-leadership.

19. Alex Beard, "China's Children Are Its Secret Weapon in the Global AI Arms Race," *Wired*, April 19, 2018, https://www.wired.co.uk/article/china-artificial -intelligence-education-superpower.

20. Diana Gehlhaus, "The Reality of America's AI Talent Shortages," *The Hill*, April 9, 2021, https://thehill.com/opinion/technology/547418-the-reality-of-americas-ai -talent-shortages; Eric Schmidt (chair) and Robert Work (vice chair), *Final Report*, National Security Commission on Artificial Intelligence, March 1, 2021, https:// www.nscai.gov/wp-content/uploads/2021/03/Full-Report-Digital-1.pdf.

21. Zwetsloot et al., *Keeping Top AI Talent in the United States*.

22. Melissa Flagg and Paul Harris, *System Re-Engineering: A New Policy Framework for the American R&D System in a Changed World* (Georgetown University Center for Security and Emerging Technology, September 2020), https://cset.georgetown.edu /wp-content/uploads/CSET-System-Re-engineering.pdf.

23. Mike Corder, "Dutch Intelligence Says It's Uncovered 2 Russian Spies," *Washington Post*, December 10, 2020, https://www.washingtonpost.com/politics/courts _law/dutch-intelligence-says-its-uncovered-2-russian-spies/2020/12/10/7366bf2e -3afc-11eb-aad9-8959227280c4_story.html.

24. Andrew Imbrie and Ryan Fedasiuk, "Untangling the Web: Why the US Needs Allies to Defend against Chinese Technology Transfer," Brookings Institution, April 27, 2020, https://www.brookings.edu/research/untangling-the-web-why-the-us-needs -allies-to-defend-against-chinese-technology-transfer/; Wm. C. Hannas and Huey-Meei Chang, *China's Access to Foreign AI Technology: An Assessment* (Georgetown University Center for Security and Emerging Technology, September 2019), 12–15, https://cset.georgetown.edu/wp-content/uploads/CSET_China_Access_To_Foreign _AI_Technology.pdf; William C. Hannas and Didi Kirsten Tatlow, *China's Quest for Foreign Technology: Beyond Espionage* (New York: Routledge, 2020); Ryan Fedasiuk and Emily Weinstein, *Overseas Professionals and Technology Transfer to China* (Georgetown University Center for Security and Emerging Technology, July 21, 2020), https://cset.georgetown.edu/wp-content/uploads/CSET-Overseas-Professionals-and -Technology-Transfer-to-China.pdf.

25. Melissa Flagg and Zachary Arnold, *A New Institutional Approach to Research Security in the United States: Defending a Diverse R&D Ecosystem* (Georgetown University Center for Security and Emerging Technology, January 2021), https://cset.georgetown.edu /research/a-new-institutional-approach-to-research-security-in-the-united-states/.

26. Daniel Zhang et al., "The AI Index 2021 Annual Report," Stanford University Human-Centered AI Institute, March 2021, https://aiindex.stanford.edu/wp-content /uploads/2021/03/2021-AI-Index-Report_Master.pdf, 167–168.

27. Husanjot Chahal, Ryan Fedasiuk, and Carrick Flynn, *Messier Than Oil: Assessing Data Advantage in Military AI* (Georgetown University Center for Security and Emerging Technology, July 2020), https://cset.georgetown.edu/wp-content/uploads /Messier-than-Oil-Brief-1.pdf.

28. Schrittwieser et al., "Mastering Atari, Go, Chess and Shogi by Planning with a Learned Model."

29. Tim Hwang, *Shaping the Terrain of AI Competition* (Georgetown University Center for Security and Emerging Technology, June 2020), https://cset.georgetown.edu /wp-content/uploads/CSET-Shaping-the-Terrain-of-AI-Competition.pdf.

30. Roxanne Heston and Helen Toner, "Have Your Data and Use It Too: A Federal Initiative for Protecting Privacy while Advancing AI," Day One Project, April 26, 2020, https://www.dayoneproject.org/post/have-your-data-and-use-it-too-a-federal -initiative-for-protecting-privacy-while-advancing-ai; Andrew Trask et al., "Beyond Privacy Trade-Offs with Structured Transparency," arXiv, December 15, 2020, http://

arxiv.org/abs/2012.08347. The OpenMined effort also deserves attention for its work in this area, https://www.openmined.org/.

31. Jack Clark, "One of the persistent issues in AI policy is that computers are . . . really fast! E.g, I talked to some government agency about GPT2 for a report that came out shortly after the launch of GPT3. We need to find ways to increase rate of information exchange between tech & gov," Twitter, December 23, 2020, 10:00 a.m., https://twitter.com/jackclarksf/status/1341805893143261184?s=21.

32. This estimate is from Jason Matheny, the former director of the Intelligence Advanced Research Projects Activity, a major funder of AI research for the federal government.

33. As discussed in chapter 4, this is called neural architecture search. Elsken, Metzen, and Hutter, "Neural Architecture Search: A Survey"; Zoph and Le, "Neural Architecture Search with Reinforcement Learning."

34. Kari Paul, "Trump Housing Plan Would Make Bias by Algorithm 'Nearly Impossible to Fight,'" *Guardian*, October 23, 2019, https://www.theguardian.com/us-news /2019/oct/22/trump-housing-plan-would-make-bias-by-algorithm-nearly-impossible -to-fight.

35. Gebru et al., "Datasheets for Datasets"; Jo and Gebru, "Lessons from Archives"; Mitchell et al., "Model Cards for Model Reporting."

36. Grother, Ngan, and Hanaoka, "Face Recognition Vendor Test."

37. Kashmir Hill, "The Secretive Company That Might End Privacy as We Know It," *New York Times*, January 18, 2020, https://www.nytimes.com/2020/01/18/technology /clearview-privacy-facial-recognition.html.

38. Andrew Imbrie et al., *Agile Alliances: How the United States and Its Allies Can Deliver a Democratic Way of AI* (Georgetown University Center for Security and Emerging Technology, February 2020), https://cset.georgetown.edu/wp-content/uploads /CSET-Report-Agile-Alliances.pdf.

39. On enhancing interoperability in multinational coalitions, see Erik Lin-Greenberg, "Allies and Artificial Intelligence: Obstacles to Operations and Decision-Making," *Texas National Security Review*, March 5, 2020, https://tnsr.org/2020/03 /allies-and-artificial-intelligence-obstacles-to-operations-and-decision-making/. On research and development, see Melissa Flagg and Paul Harris, "How to Lead Innovation in a Changed World," *Issues in Science and Technology*, September 9, 2020, https://issues.org/how-to-lead-innovation-in-a-changed-world/; Melissa Flagg, *Global R&D and a New Era of Alliances* (Georgetown University Center for Security and Emerging Technology, June 2020), https://cset.georgetown.edu/wp-content/uploads /Global-RD-and-a-New-Era-of-Alliances.pdf.

40. "OECD Principles on Artificial Intelligence," OECD, accessed July 1, 2020, https://www.oecd.org/going-digital/ai/principles/.

41. Lindsay Gorman, "The U.S. Needs to Get in the Standards Game—with Like-Minded Democracies," *Lawfare*, April 2, 2020, https://www.lawfareblog.com/us-needs -get-standards-game%E2%80%94-minded-democracies; John Seaman, *China and the*

New Geopolitics of Technical Standardization (French Institute of International Relations, January 27, 2020), https://www.ifri.org/en/publications/notes-de-lifri/china-and-new-geopolitics-technical-standardization.

42. Anna Gross, Madhumita Murgia, and Yuan Yang, "Chinese Tech Groups Shaping UN Facial Recognition Standards," *Financial Times*, December 1, 2019, https://www.ft.com/content/c3555a3c-0d3e-11ea-b2d6-9bf4d1957a67; Anna Gross and Madhumita Murgia, "China Shows Its Dominance in Surveillance Technology," *Financial Times*, December 26, 2019, https://www.ft.com/content/b34d8ff8-21b4-11ea-92da-f0c92e957a96.

43. For an example of Chinese strength in the 5G standards-setting process, see Steven Levy, "Huawei, 5G, and the Man Who Conquered Noise," *Wired*, November 16, 2020, https://www.wired.com/story/huawei-5g-polar-codes-data-breakthrough/.

44. Imbrie et al., "Agile Alliances."

45. "JAIC Facilitates First-Ever International AI Dialogue for Defense," Joint AI Center, September 16, 2020, https://www.ai.mil/news_09_16_20-jaic_facilitates_first-ever_international_ai_dialogue_for_defense.html.

46. Flournoy, Haines, and Chefitz, "Building Trust through Testing."

47. For an example of this kind of work, see Fu Ying and John Allen, "Together, the U.S. And China Can Reduce the Risks from AI," *Noema*, December 17, 2020, https://www.noemamag.com/together-the-u-s-and-china-can-reduce-the-risks-from-ai/; Horowitz and Scharre, "AI and International Stability"; Andrew Imbrie and Elsa Kania, *AI Safety, Security, and Stability Among Great Powers: Options, Challenges, and Lessons Learned for Pragmatic Engagement* (Georgetown University Center for Security and Emerging Technology, December 2019), https://cset.georgetown.edu/wp-content/uploads/AI-Safety-Security-and-Stability-Among-the-Great-Powers.pdf.

48. Kurt M. Campbell and Ali Wyne, "The Growing Risk of Inadvertent Escalation Between Washington and Beijing," *Lawfare*, August 16, 2020, https://www.lawfareblog.com/growing-risk-inadvertent-escalation-between-washington-and-beijing. For more on Chinese views of AI, see Ryan Fedasiuk, *Chinese Perspectives on AI and Future Military Capabilities* (Georgetown University Center for Security and Emerging Technology, August 2020), https://cset.georgetown.edu/wp-content/uploads/CSET-Chinese-Perspectives.pdf.

49. Carrick Flynn, *Recommendations on Export Controls for Artificial Intelligence* (Georgetown University Center for Security and Emerging Technology, February 2020), https://cset.georgetown.edu/wp-content/uploads/Recommendations-on-Export-Controls-for-Artificial-Intelligence.pdf; Saif M. Khan and Carrick Flynn, *Maintaining China's Dependence on Democracies for Advanced Computer Chips* (Brookings Institution, April 2020), https://www.brookings.edu/research/maintaining-chinas-dependence-on-democracies-for-advanced-computer-chips/.

50. "Commerce Department Further Restricts Huawei Access to U.S. Technology and Adds Another 38 Affiliates to the Entity List," Department of Commerce, August 17, 2020, https://www.commerce.gov/news/press-releases/2020/08/commerce-department-further-restricts-huawei-access-us-technology-and.

51. Lauly Li and Cheng Ting-Fang, "Inside the US Campaign to Cut China out of the Tech Supply Chain," *Nikkei*, October 7, 2020, https://asia.nikkei.com/Spotlight /The-Big-Story/Inside-the-US-campaign-to-cut-China-out-of-the-tech-supply-chain.

52. Khan and Flynn, *Maintaining China's Dependence on Democracies for Advanced Computer Chips*; Webb, "China's $150 Billion Chip Push Has Hit a Dutch Snag."

53. Saif M. Khan, *U.S. Semiconductor Exports to China: Current Policies and Trends* (Georgetown University Center for Security and Emerging Technology, October 2020), https://cset.georgetown.edu/wp-content/uploads/U.S.-Semiconductor-Exports -to-China-Current-Policies-and-Trends.pdf; CSIS Technology and Intelligence Task Force, *Maintaining the Intelligence Edge: Reimagining and Reinventing Intelligence through Innovation* (Center for Strategic and International Studies, January 2021), https:// csis-website-prod.s3.amazonaws.com/s3fs-public/publication/210113_Intelligence _Edge.pdf.

54. For examples of some discussions on this point, see Laura Rosenberger, "Making Cyberspace Safe for Democracy: The New Landscape of Information Competition," *Foreign Affairs* 99 (2020): 146; Aaron Huang, "Chinese Disinformation Is Ascendant. Taiwan Shows How We Can Defeat It," *Washington Post*, August 10, 2020, https:// www.washingtonpost.com/opinions/2020/08/10/chinese-disinformation-is -ascendant-taiwan-shows-how-we-can-defeat-it/.

55. Andrew Imbrie, Elsa Kania, and Lorand Laskai, *The Question of Comparative Advantage in Artificial Intelligence: Enduring Strengths and Emerging Challenges for the United States* (Georgetown University Center for Security and Emerging Technology, January 2020), https://cset.georgetown.edu/wp-content/uploads/CSET-The -Question-of-Comparative-Advantage-in-Artificial-Intelligence-1.pdf; Andrew Imbrie, *Competitive Strategies for Democracy in the Age of AI* (Alliance for Securing Democracy, June 30, 2020), https://securingdemocracy.gmfus.org/wp-content/uploads/2020/06 /Competitive-Strategies-for-Democracy-in-the-Age-of-AI.pdf.

INDEX

Page numbers followed by an "f" indicate figures.